T0399819

Multispecies Leisure: Human-Animal Interactions in Leisure Landscapes

Multispecies Leisure: Human-Animal Interactions in Leisure Landscapes seeks to 'bring the animal in' to the leisure studies domain and contributes to greater understanding of leisure as a complex, interwoven multispecies phenomenon.

The emerging multidisciplinary field of human-animal studies encourages researchers to move beyond narrow focus on human-centric practices and ways of being in the world, and to recognise that human and non-human beings are positioned within shared ecological, social, cultural and political spaces. With some exceptions, leisure studies has been slow to embrace the 'animal turn' and consider how leisure actions, experiences and landscapes are shaped through multispecies encounters between humans, other animals, birds and insects, plants and environment. This book begins to address this gap by presenting research that considers leisure as more-than-human experiences. The authors consider leisure *with* non-human others (e.g. dogs, horses), *affecting* those others (e.g. environmental concerns) and *affected by* the non-human (e.g. landscape, weather), by exploring the 'contact zones' between humans and other species. Thus, this work contributes to greater understanding of leisure as a complex, multispecies phenomenon.

The chapters in this book were originally published as a Special Issue of *Leisure Studies*.

Paula Danby is currently Student and Foundation Liaison Manager within the MDU and Senior Fellow of the Higher Education Academy, and was previously Lecturer in International Tourism Management at Queen Margaret University, UK. Her research focuses on human-animal relations and experiences within leisure and tourism environments, particularly equestrian tourism. Her work explores human-animal interactions for mutual well-being.

Katherine Dashper is Reader and Director of Research Degrees at Leeds Beckett University, UK. Her research applies a critical sociological lens to examine practices of work and leisure, particularly focusing on gender issues and interspecies encounters. Her multispecies research focuses mainly on human-horse interactions, and she is the author of *Human-Animal Relationships in Equestrian Sport and Leisure* (Routledge, 2017).

Rebecca Finkel is Reader and School Head of Research at Queen Margaret University, UK, and Senior Fellow of the Higher Education Academy. Main focus of research frames critical events studies within conceptualisations of social justice, equality and diversity, and cultural identity. New research explores the relational well-being dimensions of human-animal interactions in events, tourism and leisure contexts.

Multispecies Leisure: Human-Animal Interactions in Leisure Landscapes

Edited by
**Paula Danby, Katherine Dashper
and Rebecca Finkel**

LONDON AND NEW YORK

First published 2021
by Routledge
2 Park Square, Milton Park, Abingdon, Oxon, OX14 4RN

and by Routledge
52 Vanderbilt Avenue, New York, NY 10017

Routledge is an imprint of the Taylor & Francis Group, an informa business

British Library Cataloguing-in-Publication Data
A catalogue record for this book is available from the British Library

ISBN13: 978-0-367-70322-6

Typeset in Minion Pro
by codeMantra

Publisher's Note
The publisher accepts responsibility for any inconsistencies that may have arisen during the conversion of this book from journal articles to book chapters, namely the inclusion of journal terminology.

Disclaimer
Every effort has been made to contact copyright holders for their permission to reprint material in this book. The publishers would be grateful to hear from any copyright holder who is not here acknowledged and will undertake to rectify any errors or omissions in future editions of this book.

Contents

Citation Information

The chapters in this book were originally published in *Leisure Studies*, volume 38, issue 3 (June 2019). When citing this material, please use the original page numbering for each article, as follows:

For any permission-related enquiries please visit:
http://www.tandfonline.com/page/help/permissions

Contributors

Eric Brymer specialises in research that investigates the human-nature relationship with a special interest in how this relationship influences well-being and performance. Eric's expertise includes qualitative and mixed methods research design. He holds a PhD, a Master's degree in Applied Sport and Exercise Psychology and post-graduate degrees in Education and Business. He also holds research positions at Leeds Beckett University, UK; Queensland University of Technology, Australia; and the University of Cumbria, UK.

Katherine Dashper is Reader in the School of Events, Tourism and Hospitality Management at Leeds Beckett University, UK. Her research focuses on human-animal interactions and encounters. She specialises in qualitative research and is interested in exploring innovative research design to try to understand better some of the complexities of multispecies leisure and tourism.

Andrea Ford is Postdoctoral Fellow in Anthropology and the Social Sciences at the University of Chicago, USA, where she graduated with her PhD in 2017. She researches embodiment and reproduction in California. Currently, she is working on her book *Near Birth: Embodied Futures in California*, and developing a research project on reproductive toxicity and endometriosis. She has been engaged in sport riding for 20 years.

Dorothy Fox is Senior Academic in Events and Leisure Management at Bournemouth University, UK. She is the lead author of the first research methods textbook for event management students, entitled, *Doing events research: From theory to practice*. After many years of employment in business management and following completion of an undergraduate degree in environmental studies, she undertook her PhD thesis seeking to understand participation in garden visiting, within a conceptual framework of affordance theory. This approach is beginning to attract attention in leisure studies, because it enables a movement beyond the usual emphasis on individual agency to embrace social-material agents as well. Dorothy's particular area of interest is in the interactions between people and socio-natural environments.

Justin Harmon is Assistant Professor in the Department of Community and Therapeutic Recreation at the University of North Carolina Greensboro, USA. His two primary research foci are the use of music for life course development and recreation interventions post-diagnosis of cancer. When not in the classroom he is typically found in the forest with his dog or at a concert.

Eva Linghede is currently completing her PhD at the Swedish School of Sport and Health Sciences, GIH, Stockholm, Sweden, focusing on the promise of posthumanist thinking and writing in sport (studies).

Elena Marinova is First-Class Honours Bournemouth University Event Management graduate based in London. Elena completed a Marine Wildlife Guide training in the Bay of Biscay. Elena is currently working in the corporate environmental services sector, continuing her pursuit of having a positive impact on the collective thinking of, and care for the natural environment.

Vesa Markuksela is Senior Lecturer at the University of Lapland, Rovaniemi, Finland. His research interests include social theory of practice, sensory turn, affect, leisure studies, organisational research, service design and ethnographic research methods. His long-term grounding concern is the bodily and sensory encounters with human and non-human actors in the leisure Servicescape settings. He received a PhD from the University of Lapland for the thesis 'Sense like a fish – an ethnography of troll fishing brotherhoods competition practice'.

Kevin Markwell is Professor of Tourism at Southern Cross University, Australia. He has published 4 books and more than 60 book chapters and refereed journal articles on aspects of tourism and leisure relating to sexuality, nature-based tourism and human-animal relationships.

Carmel Nottle is Lecturer in Human Movement, Sport and Exercise Science at the University of South Australia. She has an interest in the human-animal intersection from a health and well-being perspective but also a particular interest in leisure activity for human companion animals and the welfare and leisure of assistance dogs.

Jeff Rose is Assistant Professor-Lecturer in Parks, Recreation and Tourism at the University of Utah, Salt Lake City, USA. His research examines systemic inequities expressed through class, race, political economy and relationships to nature. He uses this justice-focused lens on homelessness in parks, socioecological systems, outdoor education and place attachment in protected areas.

Kerry L. Sands is PhD candidate in Anthrozoology at the University of Exeter, UK. Her doctoral research is focused on exploring the lives of racing greyhounds in a Western context.

Emily Stone is PhD candidate in Anthrozoology at the University of Exeter, UK. Her doctoral research is focused on exploring the practices and discourses within pedigree cat breeding and showing in the United Kingdom.

Anu Valtonen is Professor of Cultural Economy at the University of Lapland, Rovaniemi, Finland. Her research interests relate to cultural theories and qualitative methodologies in marketing and organisational research. Her current research projects explore embodiment, sleep and senses in today's economy and society. She is also intrigued about aspects of Anthropocene and non-anthropocentric.

Jackson Wilson is Associate Professor at San Franscisco State University with research interests in the health impacts of recreation in outdoor spaces, marginalized groups in parks, and management issues in adventure education and outdoor recreation. In his roles at the university and college levels, he supports the development of institutional capacity to use online education when and how it best facilitates student success.

Janette Young is Lecturer in health promotion at the University of South Australia. She has an interest in complexity and systemic thinking particularly human–animal intersections. Her research and writing focus on understanding how relationships with animals play a role in human well-being whilst also recognising the implications for animal welfare, social and environmental justice.

Multispecies leisure: human-animal interactions in leisure landscapes

Paula Danby, Katherine Dashper and Rebecca Finkel

Introduction

Leisure is a multispecies practice. From the excitement and close interaction of human and horse tackling a show jumping course, to the joy and playfulness of a child throwing a ball for her dog, to the peaceful comfort of a human and a cat relaxing on the sofa together, leisure practices and spaces often involve multiple species, sometimes acting together, sometimes separately and sometimes in opposition. These shared and often messy entanglements between human and nonhuman animals are integral to the experiences, practices and meanings of leisure. Dashper (2018) argues that our leisure lives are often richer because of nonhuman animals, who play, relax, compete and work with and for us, and that leisure studies needs to acknowledge these more-than-human encounters if we are to understand better some of the nuances of leisure in multispecies worlds. With some exceptions (e.g. Carr, 2014; Dashper, 2017b; Markwell, 2015), leisure studies has been slow to embrace the 'animal turn' sweeping the wider social sciences and humanities, and to consider how leisure actions, experiences and landscapes are shaped through multispecies encounters between humans, other animals, birds and insects, plants and the environment. This special issue is a contribution to the project of 'bringing animals in' to leisure studies, and recognising that leisure is part of a complex, vibrant and sometimes chaotic multispecies world.

The emerging multidisciplinary field of human-animal studies encourages researchers to move beyond a narrow focus on human-centric practices and ways of being in the world, and to recognise that human and nonhuman beings are positioned within shared ecological, social, cultural and political spaces. Wider social debates related to ethics and welfare, environmental concerns and climate change, and human rights and responsibilities to the wider world, are not detached from the field of leisure studies which is both influenced by and can influence wider discourses. The broader field of human-animal studies has tended to focus on topics such as care, welfare and work, or specific human-animal encounters, such as those between people and companion animals or pets (Charles, 2014; Clarke & Knights, 2018; Coulter, 2016), and leisure has received much less focus to date. Our aim with this special issue was to challenge leisure researchers to think beyond our taken-for-granted humanist frameworks and to consider explicitly the ways in which leisure spaces and practices are co-produced, shaped and experienced by human and nonhuman animals, and what those multispecies encounters add to understandings of leisure as integral to our well-being and happiness in contemporary societies.

This introduction begins with a brief discussion of what we mean by the terms 'multispecies' and 'more-than-human' and some of the theoretical and methodological challenges that adopting posthumanist frameworks may pose for leisure researchers. We then go on to consider what such perspectives might add to the field of leisure studies, and discuss some of the existing research in this area. The next section introduces the papers in this special issue, which show the diversity and richness of multispecies perspectives on leisure, and the possibilities for advancing understanding in this emerging field. The final section suggests some areas for further development in research on multispecies leisure.

More-than-human and multispecies perspectives

Leisure studies, and the social sciences more broadly, is strongly anthropocentric, positioning humans as the only legitimate focus for study, and concentrating on human priorities, experiences and practices (Dashper, 2018; Finkel & Danby, 2018). If nonhumans do appear in research, they are usually confined to a background role, reduced to species-level, and only considered if their actions or behaviours affect human outcomes (Catlin, Hughes, Jones, Jones, & Campbell, 2013). Within this work, individual animals and their unique subjectivities disappear from view, and their 'animalness' is presented only in relation to their value to humans. A growing body of researchers are now recognising that this is untenable, and that nonhumans are more than just backdrops for human lives and are instead active agents, with their own inner lives and interests, priorities and rights (Cooke, 2011; Sanders, 1990). The seminal work of Donna Haraway (2003) has strongly influenced theoretical development in this field, and her claim that '[t]o be one is always to *become with* many' (2008,p. 4, italics in original), underpins the 'animal turn' that recognises the inseparability of human and nonhuman in what is undeniably a multispecies world. More-than-human approaches make this explicit and aim to explore the 'contact zones where lines separating nature from culture have broken down, where encounters between *Homo sapiens* and other beings generate mutual ecologies and co-produced niches' (Kirksey & Helmreich, 2010, p. 546).

More-than-human approaches within leisure aim to explore new modes of being and becoming in the contemporary world. Various theoretical approaches that rethink human-centredness have focused on the complexities surrounding interactions between humans and nonhuman animals, along with places, landscapes and objects. Posthumanism, as well as Actor Network Theory (ANT) and non-representational theories, seeks to explore and develop spatialities, politics and ethical considerations associated with humans and nonhumans, whereby the singular focus surrounding the human subject is challenged and boundaries become blurred. Instone (1998) alerts us to the fact that a postmodern world blurs the boundaries between nature, society, humans and nonhuman animals. Panelli (2010) articulates that ANT rejects the distinction between the human and the nonhuman animal, indicating that the nonhuman animal is more often than not the most important actant in the human material world. DeMello (2012) argues that nonhumans should ideally enjoy a life of love and attention, as well as humans. Bowes, Keller, Rollins, and Gifford (2015) acknowledge that trans-species social bonds are driven by multifarious factors including the desire for power, control, affection and kinship that promote wide-ranging benefits. This 'animal turn' acknowledges the embodied knowledge or indeed a 'sensorial-ontology' which arises when species meet and interact (Barad, 2008; Hayward, 2010; Hurn, 2012).

These theories and approaches are complex and diverse, and detailed discussion of them is beyond the scope of this article. However, all these positions share the posthumanist goal to decentre human authority and recognise explicitly that nonhumans can and do shape our worlds and our experiences, for good and for bad. As Peggs (2013) argues, even if we restrict our research attention to human societies and practices alone (and neither she nor we are suggesting we should do so), we still should consider relations with nonhumans in our research, as these constitute integral facets of our everyday experiences. Pacini-Ketchabaw, Taylor, and Blaise (2016, p. 2) acknowledge the long-standing resistance to accounting for the experiences and practices of nonhumans in social science research, arguing that '[t]he insistence that we live in not just exclusively human societies but in common worlds with other species runs counter to the human-centric impulse to divide ourselves off from the rest of the world and re-enact the self-perpetuating nature/culture divide.' More-than-human perspectives, in their varying forms, represent attempts to challenge this divide and recognise the complex, interwoven 'common worlds' in which we are all embedded.

As Buller (2014, p. 309) argues, 'animals are beginning, at last, to make their presence (or absence) felt and matter'. This raises challenging theoretical, methodological and practical issues for researchers. What does it mean to 'bring animals in' to research? How can we try and decentre human perspectives, and give some kind of 'voice' to nonhuman animals? How can we try and represent the deeply embodied, usually non-verbal interactions between species that constitute multispecies encounters, when we are tied by the conventions of academia to the written word? Dowling, Lloyd, and Suchet-Pearson (2017, p. 824) suggest that this needs us to radically rethink how we do research and 'to perform, to engage, to embody, to image and imagine, to witness, to sense, to analyse – across, through, with and as, more-than-humans.'

The subpractice of multispecies ethnography attempts to engage with this process, to focus on 'the lively connections among species (often, but not always, including humans), their collective effects and their ethical implications' (Pacini-Ketchabaw et al., 2016, p. 1149). This is difficult and requires attempting to shift our focus from our human perspectives alone and our preference for visual and verbal cues and, instead, to try and engage our bodies as multispecies research instruments, as 'part of the ethnographic script' (Madden, 2014, p. 282). This may encourage more interdisciplinary research in leisure studies, drawing on ethology, ecology and other natural sciences to supplement our social science perspectives and to try and begin to bridge the nature-culture divide. It may lead to more personal, introspective accounts of interspecies relationships and encounters, drawing on (auto)ethnography and narrative techniques to try and capture some of the emotive richness of multispecies leisure (see Harmon, 2019). Multispecies research has potential to disrupt dominant narratives and theoretical perspectives in leisure studies and to open up new ways of thinking about, writing about and doing research.

The more-than-human theoretical approach to human-nonhuman relations within leisure, whilst providing a more innovative mode of enquiry within the leisure landscape, also helps us to contextualise human-nonhuman experiential encounters. One of the challenges of trying to adopt a multispecies lens concerns the lack of overt descriptive reflection that arises from interspecies encounters, in that humans may find it difficult to describe, understand and explain such relationships and emotional interactions due to the lack of 'vocal' expression from nonhumans. As a result, Game (2001) argues there is a requirement for interconnectedness between species, indicating a need to respect and understand each other's differences to communicate more effectively, often on deeply embodied, nonverbal terms. She refers to an 'in-between' stage where the human becomes part nonhuman and the nonhuman part human, through sustained interaction. Including nonhuman animals as actors in research and opening up to cross-species communications emergent through the leisure landscape enables a sharing of mutual realities between humans and nonhumans (Danby, 2018). Social exchanges and embodied interaction between humans and nonhumans play significant roles, as through varying encounters human and nonhuman are able to anticipate and acknowledge each other's needs and behaviours by assessing a range of bodily cues. Such non-anthropocentric ontological perspectives emphasise how the the leisure landscape may be populated and co-constituted by varying humans and nonhumans, through myriad assemblages they engage with, together and separately (Lorimer, 2009).

To really take on multispecies perspectives is difficult, and poses challenges to leisure researchers more used to focusing on human-human interactions, and human activities, priorities and experiences. As Birke and Hockenhull (2012) articulate, studying interspecies bonds is not easy and methodologies tend to focus upon one actor rather than another, and additionally, we are dealing with relations of two very different kinds of beings. However, just because something is challenging does not mean we should not attempt to engage with it, and in the next section we introduce research on different leisure practices that draw on explicitly more-than-human perspectives, and in so doing, open up interesting theoretical, methodological and practical insights about leisure and leisure research.

Leisure as a multispecies practice

Nonhuman animals are integral to myriad human leisure experiences and help enhance many people's physical, psychological and social well-being (Danby, 2018; Dashper, 2018; Finkel & Danby, 2018; Hallberg, 2008; Young & Carr, 2018). The papers in this special issue are not the first to consider some aspects of multispecies leisure, although the earlier research is relatively dispersed around different journals and outlets. Whilst we acknowledge that this literature is diverse and covers many different contexts and issues, we have identified three core areas in the wider literature on multispecies leisure which we discuss here: dogs and dog agility; equestrian/horse leisure; and multispecies tourism. We have chosen to focus on leisure with dogs and with horses because these are the nonhuman animals with whom humans share the most intimate, active, diverse and collaborative leisure relationships. People involve both dogs and horses in a variety of leisure practices, often involving complex and nuanced interspecies communication, in ways rarely experienced between humans and other species. Multispecies tourism is our third area for discussion as it encompasses a broader variety of interspecies interactions than either human-canine or human-equine leisure, and research highlights some of the complex issues of power and responsibility that underpin all interspecies encounters, including those experienced through leisure.

Human-dog relationships are often extremely close, and offer numerous affordances for performing multispecies leisure. Carr (2014) has considered a wide range of human-dog activities and practices in his discussion of dog-related leisure, ranging from dog holidays, to dog cuisine, and dogs as 'leisure objects'. Sanders (1999) has explored human-dog interactions through various leisure and work practices, while Fletcher and Platt (2018) consider the routine dog walk as a multispecies leisure activity. Several other researchers have focused specifically on the multispecies competitive practice of dog agility, considering how involvement in this activity requires considerable investment of time, money and emotion, often placing stress on other aspects of a human's life (Baldwin & Norris, 1999; Gillespie, Leffler, & Lerner, 2002). Hultsman's (2012) research takes an interesting approach in exploring how involvement in dog agility is experienced and negotiated between human couples. Although she reported the potential for the same strains as found in previous studies, she also found that multispecies leisure provided couples with a source of close engagement and bonding, between them but also, and importantly, with their dog(s). All of these studies illustrate what many people who live with dogs (and other companion animals) know: these multispecies leisure activities are meaningful and rich expressions of complex relationships between humans and dogs, often reflective of deeply held emotions and attachments (Nottle & Young, 2019). Human-canine leisure constitutes an important part of these interspecies relationships.

There is a growing body of work that considers human-equine leisure, and although very different to that between humans and dogs, the relationship between humans and horses is also long, close and complex. Numerous equestrian practices could usefully be considered multispecies leisure, and research in this field covers multiple activities ranging from equestrian tourism (Buchmann, 2017; Dashper, 2019; Gilbert & Gillett, 2014; Sigurðardóttir & Helgadóttir, 2015) to competitive sport (Dashper, 2012; Gilbert & Gillett, 2012; Wipper, 2000) to non-competitive interactions and relationships (Dashper, 2017a; Hockenhull, Birke, & Creighton, 2010; Maurstad, Davis, & Cowles, 2013). Human-equine leisure practices and associated experiences can provide hedonistic activities and assist with the emergence and development of meaningful relationships. The 'equiscape' provides a leisure landscape through which various activities and relations are formed, where humans and horses interconnect within temporally-bound natural spaces, where boundaries become blurred (Danby, 2018; Finkel & Danby, 2018; Linghede, 2019). These and other studies explore some of the complex, deeply embodied encounters that occur between humans and horses during riding, routine interactions and caring activities (Ford, 2019; Game, 2001). Equestrian leisure requires high levels of commitment, in terms of time, emotion and financial input, and so often becomes an important marker of individual and collective identity (Dashper, 2017b; Dashper, Abbott, & Wallace, 2019). Dominant themes emerging from the human-horse leisure literature include: the gendered nature of this form of multispecies leisure (Dashper, 2016; Finkel & Danby, 2018; Linghede, 2019); the role of equestrian leisure at different stages of the (human and equine) lifecourse (Davis, Maurstad, & Dean, 2016; Franklin & Schuurman, 2017; Sanchez, 2017); and, the importance of partnership in human-horse relationships (Dashper, 2017b; Maurstad et al., 2013). This growing field of study illustrates some of the complexities of multispecies leisure, which can be simultaneously joyous and rewarding, as well as risky and potentially heartbreaking.

The therapeutic role of animals within leisure is widely acknowledged and forms a significant part of the leisure science community due to the broad ranging therapeutic, psychological and physical health benefits associated with such human-animal encounters (Chandler, 2012; Fine, 2015; Krause-Parello, Gulick, & Basin, 2019; Nimer & Lundahl, 2007) and, as a result, multifarious animal assisted therapy (AAT) interventions (particularly with the inclusion of dogs and horses) have been incorporated into diverse social practices to improve emotional and physical wellbeing. As Chandler (2012, p. xi) states, 'Humans are designed to thrive through relationships. Our emotional, physical and spiritual essence craves connections with others, not only so that we may have our needs met out so that we may also experience purpose and meaning in the time that we dwell on this Earth.' It may be argued that this raises ethical considerations and requires specific regulations to promote and perform successful leisure-led AAT for both the wellbeing of humans and nonhumans.

Relationships between humans and dogs and between humans and horses differ in important ways that reflect the different ways in which we live, communicate and interact with different species and with different individual animals. Leisure with dogs and with horses offers people many opportunities for rewarding (and indeed sometimes challenging) interspecies encounters, and for developing and maintaining close interspecies relationships, and tends to reflect deep commitment from human participants in relation to time, money and emotion (Dashper et al., 2019). However, multispecies interactions are always underpinned by complex power relations as both human and nonhuman are positioned in a human-centric world that prioritises human interests over nonhuman ones (Carter & Charles, 2013). Whether they are called 'pets', 'partners' or 'collaborators', dogs, horses and all animals we actively involve in our leisure activities have not actively chosen to do so in the way as we as humans decide how to spend our free time. This raises complex questions about the morality of involving nonhumans in our leisure practices, and the responsibilities we owe to them if we do. A few studies have started to address these issues in relation to 'pets' and other companion species, but this has yet to be considered fully in relation to leisure studies, a point to which we return below (Dashper, 2014; Irvine, 2004).

Tourism offers another important focus for the emerging field of multispecies leisure. While there is a wide range of research exploring different aspects of wildlife and ecotourism (Curtin & Kragh, 2014; Reynolds & Braithwaite, 2001), much of this does not take what we would consider a multispecies or more-than-human perspective, and focuses very much on human interests, experiences and practices, with nonhuman animals featuring as attractions, or as part of the background to human activities and associated entertainment. In contrast, Actor Network Theory has proven popular within tourism studies, and has been used to consider more-than-human aspects of situations as diverse as the interface between science and wildlife tourism (Rodger, Moore, & Newsome, 2009), cheese as a local tourism actor (Ren, 2011) and actor-networks in gorilla tourism (van der Duim, Ampumuza, & Ahebwa, 2014). Warkentin's (2010, 2011) research on swimming with dolphins argues for the importance of what she terms 'interspecies etiquette' in these multispecies tourism encounters, which form memorable and unusual leisure experiences for the human participant, but have potential to be distressing for the dolphins. Bertella (2014, p. 122) argues that nonhuman animals should be included as 'central actors in the tourism network', and we agree that tourism offers many interesting avenues for exploring different aspects of interspecies interactions as is evident in Markwell's (2015) edited collection, and an important site of multispecies leisure.

Tourism offers many humans an opportunity to see and interact with diverse species on a global stage that we normally would not encounter, often in their own natural environments. This can be exciting for tourists and may contribute to conservation efforts through better interspecies understanding and awareness of the importance and diverse needs of other species, which could be particularly vital as we face challenges to do with human-induced climate change and other sustainability concerns. However, tourism involving other animals – whether they be captive animals (e.g. in zoos or parks) or in 'natural' environments – raises many difficult questions about animal welfare, human impacts on other species and their environments, and the ethics and responsible behaviours of animal-related tourism (Carr & Broom, 2018; Fennell, 2011). We return to some of these issues further in the final section.

These, and other studies, illustrate some of the diversity of research on multispecies leisure, in theoretical and methodological terms, as well as in relation to the focus of interspecies encounters, and the typology of nonhuman animals involved, and also suggest some areas for further development and critical reflection. This is reflected within this special issue, and in the next section we introduce the papers that form the collection.

Introduction to the papers in the special issue

This special issue highlights the diverse landscape of human-nonhuman encounters in leisure. Included is work that not only focuses on pets and companion animals, such as dogs and horses, but also draws attention to less studied nonhumans, such as reptiles, fish, and coyotes. The heterogeneity of species at the centre of leisure research has led us to think in different ways about how these papers should be grouped and ordered in this special issue. In many respects, the 'type' of nonhuman animal is not the main discerning factor for the research. Instead, ontological perspectives and methodological approaches can be seen to be the innovative aspects for further-ing understanding and engagement in this subject area. A range of qualitative methods have been employed by all of the authors, which is appropriate given the exploratory nature of this kind of research which is often deeply embodied and imbued with meaning (Barad, 2007), although Wilson and Rose (2019) illustrate the contribution that quantitative and mixed methods approaches can bring to understandings of multispecies leisure. The way that researchers reflect, observe, analyse, and recount the various lived experiences of *being with* nonhuman animals (Haraway, 2003) is at the core of understanding multispecies leisure. Therefore, we have set out the articles based on methodological approaches.

The special issue begins with autoethnographic approaches to researching multispecies leisure. Although these accounts are conveyed from a human point of view, all of the authors took into consideration nonhuman perspectives to prioritise co-creation of research. In a departure from traditional human-centric leisure studies, the papers explore the personal lived experiences of interspecies encounters. Nottle and Young (2019) consider the intersection of animal leisure with human leisure in a reflective analysis of individual human-nonhuman animal preferences and personalities. Focusing on different approaches to 'fur parenting' and leisure lives, the authors observe their lives with their five dogs, framed within Stebbins' Serious Leisure Perspective categorisation (Elkington & Stebbins, 2014). Harmon (2019) also conducts research with dogs, but situated within end of life contexts. By studying meaningful multispecies relationships with regard to mortality, he considers the therapeutic qualities of human-dog interactions in nature. The auto/ethnography approach enables Harmon (2019) to express some of the affective and deeply held emotional aspects of multispecies encounters. Next, Ford (2019) presents an auto-ethnography of the experience of sport horse riding, with emphasis on co-embodiment between horse and human. Drawing on decades of personal experience in sport horse riding, she builds upon phenomenological and anthropological theories of embodiment. Following this, Markwell (2019) contemplates his life-long interest in amateur herpetology. His analysis of the intersections between reptiles, leisure, place and identity within his own life experience reveals how multi-species leisure can lead to increased empathy and understanding across species boundaries. All of the authors in this section recognise the relational and emotional capabilities of interspecies interactions (Gruen, 2011) and the mutual satisfaction that can be gained from leisure experiences with one another.

Moving from autoethnography to ethnography, the following papers share accounts of participant and direct observation of multispecies leisure in a range of settings. Markuksela and Valtonen (2019) provide insights into the rhythmic nature of the waterscape with their exploration of human-nonhuman encounters in the leisure activity of match fishing. Based on findings from three-year sensory ethnographic fieldwork conducted in Finnish Lapland, the paper suggests that these kinds of human-nonhuman encounters can be characterised as a dance between a fish and an angler. Moving to the streets of Wales, Sands (2019) investigates the affective spaces of charitable human-greyhound gatherings, which prompt further emotional, economic and prac-tical exchanges. Her study reminds us of some of the ethical dimensions of multispecies leisure practices. Next, in contrast to the numerous studies on dog shows and agility discussed above, Stone (2019) examines the breeding and showing of pedigree cats, with an emphasis on cats' perspectives. She argues that there is a need for concentration on more equal, mutual wellbeing in such leisure environments, as there is currently a favouring of human experiences in cat shows. Dashper and Brymer (2019) introduce an ecological-phenomenological framework for under-standing relationships between animate actors and their environment in and through leisure, by using the example of human riders and horses in the context of a pleasure ride leisure event. They underline the importance of considering all practices and interactions in relation to the environ-ments in which they take place. All of the ethnographic accounts in this special issue can be seen to progress Haraway's (2003) idea of *naturecultures*, which suggests that nature and culture are not oppositions. By recognising the mutual benefits for human and nonhuman animals, the authors advocate new ways of thinking about and investigating the value and meaning of leisure in multispecies contexts.

The next set of papers draw upon interview and dialogue techniques. Linghede (2019) explores human-horse relations and intersectionality in boys' equestrian stories, through the concept of intractivity and creative analytical writing. She found that engaging with horses can encourage boys to be less constrained by dominant gender discourses and that transcending the human-animal divide can help them to transcend the female-male/masculine-feminine divide. Following this, Marinova and Fox (2019) analyse ethical issues related to animal rights and welfare in planned event environments, and the uncertainty regarding animals' status as stakeholders during live events. They found that Millennials are concerned about animal welfare, although this is underpinned by ambiguity and contradiction in relation to the involvement of different animals in different types of events. Lastly, utilising a mixed methods approach, Wilson and Rose (2019) investigate the preferences of people in the United States for sharing leisure space in their local urban parks with coyotes, a nonhuman animal that often provokes more negative responses from people than many of the others that feature within this special issue. They found that coyotes are perceived as dangerous, and, thus, people are unwilling to share leisure spaces with them, which has implications for current efforts to promote human-wildlife coexistence strategies in many urban locations.

Throughout these studies, current debates and ongoing discussions about human-nonhuman animal coexistence and interactions in leisure landscapes are seen to be complex, affective, experiential and intractive. Therefore, this special issue contributes to greater understanding of leisure as an entangled multispecies phenomenon by presenting international scholarship utilising creative methodological approaches which explore a range of issues and perspectives on multispecies leisure in order to contribute to the development of critical leisure theory and human-nonhuman animal studies.

Future development for multispecies leisure research

This special issue is a contribution to the development of multispecies insights on leisure. This research area remains relatively emergent and has potential for further development and sophistication in order to enhance theorising in leisure studies and also to contribute to theoretical, methodological and practical developments in the broader field of human-animal studies. In this final section, we make some suggestions for future development.

Many of the papers in this special issue, and numerous others in the wider literature, focus predominantly on the positive, beneficial aspects of multispecies encounters and leisure and reflect close bonds – even love – across species boundaries (Nottle & Young, 2019). Nickie Charles' research has posited that pets can be understood to be kin – family members in many circumstances – and that they provide emotional support, comfort and security to many people (Charles, 2014; Charles & Davies, 2011) and often serve as substitutes for human relations. Leisure researchers could usefully add to this line of enquiry, considering if and how pets (and other animals in some circumstances) are incorporated as family members into family leisure activities. Multispecies families are a reality for many and considering if and how the more-than-human aspects of family leisure help maintain familial bonds, at times contributing to overcoming tension and sometimes leading to problems and conflict, would advance understanding of multispecies families and leisure.

Whilst many multispecies interactions are positive and based on genuine affection and respect, we would caution against overly-romanticising interspecies relationships. Even pets, with whom many share their homes and everyday lives, are in a liminal position relative to the broader family; often considered full family members and valued for their 'animalness', but still subject to human whim and attempts to 'civilise' their behaviour through practices like selective breeding, training and neutering (Fox, 2006). Pets and other companion species, like horses, are still classed as human property and are liable to be sold or euphemistically 'destroyed' if they do not live up to human expectations, behave in a way deemed unacceptable to their human owners or simply become surplus to human wants and requirements (Dashper, 2014; McCarthy, 2016). Critical examinations of power within multispecies leisure practices could usefully address some of these issues and consider what we owe to the animals we involve in our leisure practices, including the responsibilities we have to them and the potential for abusing that power.

There are also many examples where nonhuman animals are involved and in many situations exploited within human leisure in ways that are clearly to the detriment of those animals, from hunting, to fighting, to exhibition in zoos and parks with low standards of animal welfare. There is research in the tourism field that considers some of these issues (Fennell, 2013), and this can be developed further to explore what we mean by 'good welfare' in the context of global tourism. Further, attitudes to nonhuman animals are historically, culturally and socially specific, and we would like to see further examination of multispecies leisure in different spaces, cultures and societies. This may lead to serious consideration of what might be considered to be ethical or morally acceptable ways in which to engage nonhuman animals in our leisure practices in order to try to respect nonhuman subjectivity while maintaining human pleasure in these activities.

These discussions are important in terms of our relationships with and attitudes to the nonhuman world, issues which are particularly pertinent as we face the potentially catastrophic consequences of human effects on the climate and nature. As Birke (2007, p. 306, italics in original) argues, 'nonhuman animals matter for *themselves*' and so we have a moral obligation to critically reflect on multispecies leisure and how our pleasure can affect other animals. At the same time, multispecies perspectives are useful for helping us understand the human world better as well, particularly in relation to oppression, exploitation and inequality. Birke (2007, p. 307) argues that:

> [E]ach of the ways in which 'othering' appears in our culture is mutually reinforcing. Sexism, racism, imperialism, and our treatment of nonhuman animals are all deeply interrelated and deeply entwined.

Consequently, multispecies perspectives on leisure can help contribute to understanding of human inequalities. As Nibert (2003) argues, how we treat and think about nonhuman animals is often caught up with what happens to many humans.

Multispecies perspectives on leisure thus have potential to advance understanding of both inter-species interactions *and* human-based systems of inequality in and through leisure. The so-called 'animal turn' is beginning to be felt within leisure studies, and we believe this opens up many fruitful avenues for critical reflection on leisure and its roles and influences within our multi-species worlds.

Disclosure statement

No potential conflict of interest was reported by the authors.

References

Baldwin, C. K., & Norris, P. A. (1999). Exploring the dimensions of serious leisure: "Love me—Love my dog!". *Journal of Leisure Research*, 31(1), 1–17.

Barad, K. (2007). *Meeting the universe halfway: Quantum physics and the entanglement of matter and meaning.* Durham, NC: Duke University Press.

Barad, K. (2008). Queer causation and the ethics of mattering. In N. Giffney & M. J. Hird (Eds.), *Queering the non-human.* Aldershot: Ashgate.

Bertella, G. (2014). The co-creation of animal-based tourism experience. *Tourism Recreation Research*, 39(1), 115–125.

Birke, L. (2007). Relating animals: Feminism and our connections with nonhumans. *Humanity & Society*, 31(4), 305–318.

Birke, L., & Hockenhull, J. (2012). Investigating human-animal bonds: Realities, relating, research. In L. Birke & J. Hockenhull (Eds.), p. 15-36. *Crossing boundaries. Investigating human-animal relationships.* Leiden: Koninklijke Brill.

Bowes, M., Keller, P., Rollins, R., & Gifford, R. (2015). Parks, dogs, and beaches: Human- wildlife conflict and the politics of place. In N. Carr (Ed.), p. 146-174. *Domestic animals and leisure: Leisure studies in a global era.* London: Palgrave Macmillan.

Buchmann, A. (2017). Insights into domestic horse tourism: The case study of Lake Macquarie, NSW, Australia. *Current Issues in Tourism, 20*(3), 261–277.

Buller, H. (2014). Animal geographies I. *Progress in Human Geography, 38*(2), 308–318.

Carr, N. (2014). *Dogs in the leisure experience.* London: CABI.

Carr, N., & Broom, D. M. (2018). *Tourism and animal welfare.* London: CABI.

Carter, B, & Charles, N. (2013). Animals, agency and resistance. *Journal for The Theory Of Social Behaviour, 43*(3), 322–340. doi:10.1111/jtsb.2013.43.issue-3

Catlin, J., Hughes, M., Jones, T., Jones, R., & Campbell, R. (2013). Valuing individual animals through tourism: Science or speculation? *Biological Conservation, 157,* 93–98.

Chandler, C. (2012). *Animal assisted therapy in counseling.* Abingdon: Routledge.

Charles, N. (2014). 'Animals just love you as you are': Experiencing kinship across the species barrier. *Sociology, 48* (4), 715–730.

Charles, N., & Davies, C. A. (2011). My family and other animals: Pets as kin. *Sociological Research Online, 13*(5), 4.

Clarke, C., & Knights, D. (2018). Who's a good boy then? Anthropocentric masculinities in veterinary practice. *Gender, Work & Organization.* doi:10.1111/gwao.12244

Cooke, S. (2011). Duties to companion animals. *Res Publica, 17*(3), 261.

Coulter, K. (2016). Beyond human to humane: A multispecies analysis of care work, its repression, and its potential. *Studies in Social Justice, 10*(2), 199–219.

Curtin, S., & Kragh, G. (2014). Wildlife tourism: Reconnecting people with nature. *Human Dimensions of Wildlife, 19*(6), 545–554.

Danby, P. (2018). Post-humanistic insight into human-equine interactions and wellbeing within leisure and tourism. In J. Young & N. Carr (Eds.), *Domestic animals, humans, and leisure* (pp. 58–176). Abingdon: Routledge.

Dashper, K. (2012). Together, yet still not equal? Sex integration in equestrian sport. *Asia-Pacific Journal of Health, Sport and Physical Education, 3*(3), 213–225.

Dashper, K. (2014). Tools of the trade or part of the family? Horses in competitive equestrian sport. *Society & Animals, 22*(4), 352–371.

Dashper, K. (2016). Strong, active women: (Re)doing rural femininity through equestrian sport and leisure. *Ethnography, 17*(3), 350–368.

Dashper, K. (2017a). Listening to horses: Developing attentive interspecies relationships through sport and leisure. *Society & Animals, 25*(3), 207–224.

Dashper, K. (2017b). *Human-animal relationships in equestrian sport and leisure.* Abingdon: Routledge.

Dashper, K. (2018). Moving beyond anthropocentrism in leisure research: Multispecies perspectives. *Annals of Leisure Research, 22*(2), 133–139.

Dashper, K. (2019). More-than-human emotions: Multispecies emotional labour in the tourism industry. *Gender, Work & Organization.* doi:10.1111/gwao.12344

Dashper, K., Abbott, J., & Wallace, C. (2019). 'Do horses cause divorces?' Autoethnographic insights on family, relationships and resource intensive leisure. *Annals of Leisure Research,* 1–18. doi:10.1080/11745398.2019.1616573

Dashper, K., & Brymer, E. (2019). An ecological-phenomenological perspective on multispecies leisure and the horse-human relationship in events. *Leisure Studies,* 1–14. doi:10.1080/02614367.2019.1586981

Davis, D. L., Maurstad, A., & Dean, S. (2016). 'I'd rather wear out than rust out': Autobiologies of ageing equestriennes. *Ageing & Society, 36*(2), 333–355.

DeMello, M. (2012). *Animals and society: An introduction to human-animal studies.* New York: Columbia University Press.

Dowling, R., Lloyd, K., & Suchet-Pearson, S. (2017). Qualitative methods II: 'More-than-human' methodologies and/in praxis. *Progress in Human Geography, 41*(6), 823–831.

Elkington, S., & Stebbins, R. (2014). *The serious leisure perspective: An introduction.* Abingdon: Routledge.

Fennell, D. A. (2011). *Tourism and animal ethics.* Abingdon: Routledge.

Fennell, D. A. (2013). Tourism and animal welfare. *Tourism Recreation Research, 38*(3), 325–340.

Fine, A. (2015). *Handbook on animal-assisted therapy: Foundations and guidelines for animal assisted interventions.* London: Elsevier.

Finkel, R., & Danby, P. (2018). Legitimising leisure experiences as emotional work: A post-humanist approach to gendered equine encounters. *Gender, Work and Organization.* doi:10.1111/gwao.12268

Fletcher, T., & Platt, L. (2018). (Just) a walk with the dog? Animal geographies and negotiating walking spaces. *Social & Cultural Geography, 19*(2), 211–229.

Ford, A. (2019). Sport horse leisure and the phenomenology of interspecies embodiment. *Leisure Studies,* 1–12. doi:10.1080/02614367.2019.1584231

Fox, R. (2006). Animal behaviours, post-human lives: Everyday negotiations of the animal–Human divide in pet-keeping. *Social & Cultural Geography*, 7(4), 525–537.

Franklin, A., & Schuurman, N. (2017). Aging animal bodies: Horse retirement yards as relational spaces of liminality, dwelling and negotiation. *Social & Cultural Geography*, 1–20.

Game, A. (2001). Riding: Embodying the centaur. *Body & Society*, 7(4), 1–12.

Gilbert, M., & Gillett, J. (2012). Equine athletes and interspecies sport. *International Review for the Sociology of Sport*, 47(5), 632–643.

Gilbert, M., & Gillett, J. (2014). Into the mountains and across the country: Emergent forms of equine adventure leisure in Canada. *Loisir Et Société/Society and Leisure*, 37(2), 313–325.

Gillespie, D. L., Leffler, A., & Lerner, E. (2002). If it weren't for my hobby, I'd have a life: Dog sports, serious leisure, and boundary negotiations. *Leisure Studies*, 21(3–4), 285–304.

Gruen, L. (2011). *Ethics and animals. An introduction.* Cambridge: Cambridge University Press.

Hallberg, L. (2008). *Walking the way of the horse: Exploring the power of the horse-human relationship.* New York: iUniverse.

Haraway, D. (2003). *The companion species manifesto: Dogs, people and significant otherness.* Chicago: Prickly Paradigm Press.

Haraway, D. (2008). *When species meet.* Minneapolis: University of Minnesota Press.

Harmon, J. (2019). Tuesdays with Worry: Appreciating nature with a dog at the end of life. *Leisure Studies*. doi:10.1080/02614367.2018.1534135

Hayward, E. (2010). Fingereyes: Impressions of cup corals. *Cultural Anthropology*, 25(4), 577–599.

Hockenhull, J., Birke, L., & Creighton, E. (2010). The horse's tale: Narratives of caring for/about horses. *Society & Animals*, 18(4), 331–347.

Hultsman, W.Z. (2012). Couple involvement in serious leisure: examining participation in dog agility. *Leisure Studies*, 31(2), 231-253. doi:10.1080/02614367.2011.619010

Hurn, S. (2012). *Humans and other animals: Cross cultural perspectives on human-animal interactions.* London: Pluto Press.

Instone, L. (1998). The coyote's at the door: Revisioning human-environment relations in the Australian context. *Cultural Geographies*, 5(4), 452–467.

Irvine, L. (2004). Pampered or enslaved? The moral dilemmas of pets. *International Journal of Sociology and Social Policy*, 24(9), 5–17.

Kirksey, S. E., & Helmreich, S. (2010). The emergence of multispecies ethnography. *Cultural Anthropology*, 25(4), 545–576.

Krause-Parello, C., Gulick, E., & Basin, B. (2019). Loneliness, depression and physical activity in older adults: The therapeutic role of human-animal interactions. *Anthrozoos*, 32(2), 239–254.

Linghede, E. (2019). Becoming horseboy(s) – Human-horse relations and intersectionality in equiscapes. *Leisure Studies*, 1–14. doi:10.1080/02614367.2019.1584230

Lorimer, H. (2009). Posthumanism/posthumanistic geographies. *International Encyclopedia of Human Geographies*, 8, 344–354.

Madden, R. (2014). Animals and the limits of ethnography. *Anthrozoös*, 27(2), 279–293.

Marinova, E., & Fox, D. (2019). An exploratory study of British Millennials' attitudes to the use of live animals in events. *Leisure Studies*, 1–13. doi:10.1080/02614367.2019.1583766

Markuksela, V., & Valtonen, A. (2019). Dance with a fish? Sensory human-nonhuman encounters in the waterscape of match fishing. *Leisure Studies*, 1–14. doi:10.1080/02614367.2019.1588353

Markwell, K. (2015). Birds, beasts and tourists: Human-animal relationships in tourism. In K. Markwell (Ed.), *Animals and tourism: Understanding diverse relationships* (pp. 1–23). Bristol: Channel View Publications.

Markwell, K. (2019). Relating to reptiles: An autoethnographic account of animal–Leisure relationships. *Leisure Studies*. doi:10.1080/02614367.2018.1544657

Maurstad, A., Davis, D., & Cowles, S. (2013). Co-being and intra-action in horse–Human relationships: A multispecies ethnography of be(com)ing human and be(com)ing horse. *Social Anthropology*, 21(3), 322–335.

McCarthy, D. (2016). Dangerous dogs, dangerous owners and the waste management of an 'irredeemable species'. *Sociology*, 50(3), 560–575.

Nibert, D. (2003). Humans and other animals: Sociology's moral and intellectual challenge. *International Journal of Sociology and Social Policy*, 23(3), 4–25.

Nimer, J., & Lundahl, B. (2007). Animal-assisted therapy: Meta-analysis. *Anthrozoos*, 20(3), 225–238.

Nottle, C., & Young, J. (2019). Individuals, instinct and moralities: Exploring multi-species leisure using the serious leisure perspective. *Leisure Studies*, 1–14. doi:10.1080/02614367.2019.1572777

Pacini-Ketchabaw, V., Taylor, A., & Blaise, M. (2016). Decentring the human in multispecies ethnographies. In C. Taylor & C. Hughes (Eds.), *Posthuman research practices in education* (pp. 149–167). London: Palgrave Macmillan.

Panelli, R. (2010). more-than-human social geographies: posthuman and other possibilities. *Progress in Human Geography*, 34(1), 79-87. doi:10.1177/0309132509105007

Peggs, K. (2013). The 'animal-advocacy agenda': Exploring sociology for non-human animals. *The Sociological Review, 61*(3), 591–606.

Ren, C. (2011). Non-human agency, radical ontology and tourism realities. *Annals of Tourism Research, 38*(3), 858–881.

Reynolds, P. C., & Braithwaite, D. (2001). Towards a conceptual framework for wildlife tourism. *Tourism Management, 22*(1), 31–42.

Rodger, K., Moore, S. A., & Newsome, D. (2009). Wildlife tourism, science and actor network theory. *Annals of Tourism Research, 36*(4), 645–666.

Sanchez, L. (2017). 'Every time they ride, I pray:' Parents' management of daughters' horseback riding risks. *Sociology of Sport Journal, 34*(3), 259–269.

Sanders, C. (1990). The animal 'other': Self-definition, social identity and companion animals. *Advances in Consumer Research, 17*, 662–668.

Sanders, C. (1999). *Understanding dogs.* Temple: Temple University Press.

Sands, K. (2019). Shared spaces on the street: A multispecies ethnography of ex-racing greyhound street collections in South Wales, UK. *Leisure Studies*, 1–14. doi:10.1080/02614367.2019.1577904

Sigurðardóttir, I., & Helgadóttir, G. (2015). Riding high: Quality and customer satisfaction in equestrian tourism in Iceland. *Scandinavian Journal of Hospitality and Tourism, 15*(1–2), 105–121.

Stone, E. (2019). What's in it for the cats?: Cat shows as serious leisure from a multispecies perspective. *Leisure Studies*, 1–13. doi:10.1080/02614367.2019.1572776

van der Duim, R., Ampumuza, C., & Ahebwa, W. M. (2014). Gorilla tourism in Bwindi Impenetrable National Park, Uganda: An actor-network perspective. *Society & Natural Resources, 27*(6), 588–601.

Warkentin, T. (2010). Interspecies etiquette: An ethics of paying attention to animals. *Ethics & the Environment, 15*(1), 101–121.

Warkentin, T. (2011). Interspecies etiquette in place: Ethical affordances in swim-with-dolphins programs. *Ethics & the Environment, 16*(1), 99–122.

Wilson, J., & Rose, J. (2019). A predator in the park: Mixed methods analysis of user preference for coyotes in urban parks. *Leisure Studies*, 1–17. doi:10.1080/02614367.2019.1586979

Wipper, A. (2000). The partnership: The horse-rider relationship in eventing. *Symbolic Interaction, 23*(1), 47–70.

Young, J., & Carr, N. (2018). *Domestic animals, humans, and leisure: Rights, welfare, and wellbeing.* Abingdon: Routledge.

Individuals, instinct and moralities: exploring multi-species leisure using the serious leisure perspective

Carmel Nottle🆔 and Janette Young🆔

ABSTRACT

While there has been increasing interest in the human-animal leisure intersection in recent times, leisure still largely remains human-centric or focussed. Much remains to be explored in seeking to understand animal leisure, and the intersection of animal leisure with human leisure. Spring boarding from Franklin's argument that understanding cross species involvement calls for intense, reflective analyses that can begin in our own human lives and experiences we use an ethnographic approach to explore the intersection of human and animal leisure's. Use is made of Stebbins' Serious Leisure Perspective (SLP) categorisation to present observations, analyses and learnings as to the leisure lives of the 5 dogs that share our (the authors) very different multi-species leisure lives. Our explorations demonstrate that multi-species leisure cannot be presumed; and that experiences of leisure per se intersect with individual animal preferences and personalities. We also identify the potential to see some animal leisure as Serious-Amateur and even Devotee Work when incorporating 'instinct' (the outcome of generational human control of some species fertility) into considerations, and briefly explore the extension of human parenting and leisure moralising to the rising profile of fur-parenting.

Introduction

As a field, leisure studies has a history of recognising the inclusion of animals in human leisure (Graham & Glover, 2014; Saunders & Turner, 1987; Siderelis, 2001). In more recent times the welfare and wellbeing of animals integral to human leisure has also been receiving attention, this has included exploring the needs of animals for leisure, and the development of shared leisure places and psyches (Danby, 2018). Enquiries that emerge from these explorations include delving more deeply into questions regarding *animal* leisure. The three core questions explored here are: What does animal leisure look like and how can we understand it? Does human leisure with animals necessarily equate to animal leisure, and vice versa? What may be nuances of animal leisure that we as humans need to recognise?

While perhaps ultimately these discussions can only be human-centric, interests in animal sentience and rights, and shifts in how animals are being perceived in the twenty first century necessitates considerations of animal leisure. Changes include 'fur-babies' becoming increasingly recognised members of human families with a boom in doggie day care, pet hotels, pet-focussed mega-stores and dramatically increased proportions of individual spending on pets (Animal

🔗 Supplemental data for this article can be accessed here.

Medicines Australia, 2016). While many of these societal changes are yet to be embedded in public policy and legal frameworks even this is shifting, for example in the UK police dogs now receive 'pensions' (Cochrane, 2016). All these changes in addition to the animal-turn across multiple academic fields (religion, politics, leisure) indicates that the questions explored here have increasing legitimacy.

Methodology

In this multi-ethnographic research we use the two very different canine-human families of our own (the authors) lives to seek to understand animal leisure. Carmel is highly engaged with her dogs in formal dog sports and associated activities; Janette's dogs live a very domestic life with no formal activities. While our approach can be critiqued as continuing the privileging of human-canine relationships (Young & Carr, 2018), we believe that conceptualising animal leisure is a serious rights matter. It will be for future researchers to extend our analyses and thinking to other species. Our aim is to initiate discussions that conceptualise animal leisure(s) and particularly domestic/companion animal's leisure. These animals can be seen as paradoxically privileged and extremely vulnerable (Young & Carr, 2018). Understanding what leisure is for these animals as sentient others is core to recognising animals as fellow members of the Anthropocene.

Historically ethnography has focussed on seeking to understand cultures of groups to whom the researcher is an outsider using careful observation of social actors as they live their everyday lives. Such observations are then combined with other sources to build thick rich descriptions (Geertz, 1973; Gobo, 2008). Franklin (2015) argued that understanding cross species involvement calls for intense, reflective analyses that begins in our own human lives and experiences using *auto*ethnography. Madden (2014) has suggested that animals cannot engage in ethnographic research due to their lack of language and hence inability to convey their thinking (if one considers that they think). We would argue that this is a limited understanding of communication. As health professionals both of us have worked with non-verbal human peers and are conscious of the need to develop acute observational skills in order to maximise communication in the absence of language. An absence that may be accompanied by cognitive/thinking impacts (eg stroke, head injury, dementia). Secondly, as any pet owner can attest, animals clearly communicate, with each other and across species.

Our analysis is underpinned by the perspective that both humans and animals have individual sentience; that is the capacity to subjectively experience individually identifiable experiences of leisure (Dawkins, 2006; Young & Carr, 2018). This paper is predominately ethnographic as we observe our canine companions closely seeking to understand their leisure. We recognise that their leisure is in fact constrained by the inequitable power relationship that exists in the human-centric world that all seven of us live in. But we wanted to position our dogs on the main-stage of this paper, with reflective insights into us (the human members of our respective packs) where we believe these are pertinent to understanding canine leisure. We recognise that we could be criticised for presenting a relatively thin autoethnographic approach, but our passion is to seek to conceptualise our non-human friends leisure, however imperfectly.

The serious leisure perspective (SLP)

Stebbins has been developing the Serious Leisure Perspective (SLP) framework for over 40 years (Stebbins, 2017) and while not without critiques (Veal, 2017) it has been used extensively as a framework for understanding human leisure (Hultsman, 2012; Scott & McMahan, 2017; Yang, Kim, & Heo, 2018). We have used the SLP to provide an exploratory and organising framework to consider animal leisure here. This provides the opportunity for future explorations that compare our analysis with others (both human and animal) using the SLP. The SLP is composed of a range of categories and levels grouped into three main sub-categories; Casual and Project Based Leisure, and

Serious Pursuits. The meanings of categories are outlined as we make use of them here. Stebbins also theorizes that probably all leisure can be engaged in to varying levels – from Neophyte (dabbling in an activity) to Devotee worker (Scott & McMahan, 2017; Stebbins, 2013). We use both these aspects of the SLP to explore notions of multi (or bi-) species leisure(s).

Figure 1 is a diagrammatic representation of the category tree of SLP (Hartel, 2013) with the SLP categories of leisure that we discern in our analysis highlighted. These categories are used to structure the main body of this paper. The supplemental online materials linked to this paper give readers a visual insight into the analysis presented here.

Two multi-species households

To begin to explore the leisure of our canine companion's lives the following section provides a snapshot of our two diverse canine-human leisure lives. Our interest in understanding the leisure of our canine family members emerges out of six years of conversations, sharing of photos and Facebook postings, and more recently academic explorations (Young & Nottle, 2017). Developing and writing this paper has allowed us to systematically reflect, document and analyse an aspect of what has been everyday cultural life.

Carmel, gypsy and Bunji

Carmel, the human member of our first co-species team works fulltime as an academic at an Australian university. In 2010 she acquired Gypsy to be her non-human companion. After having Gypsy for a year, Carmel was concerned that Gypsy was becoming lonely and stressed on her own during the long days that Carmel needs to put in at work, away from home in a locale that Gypsy

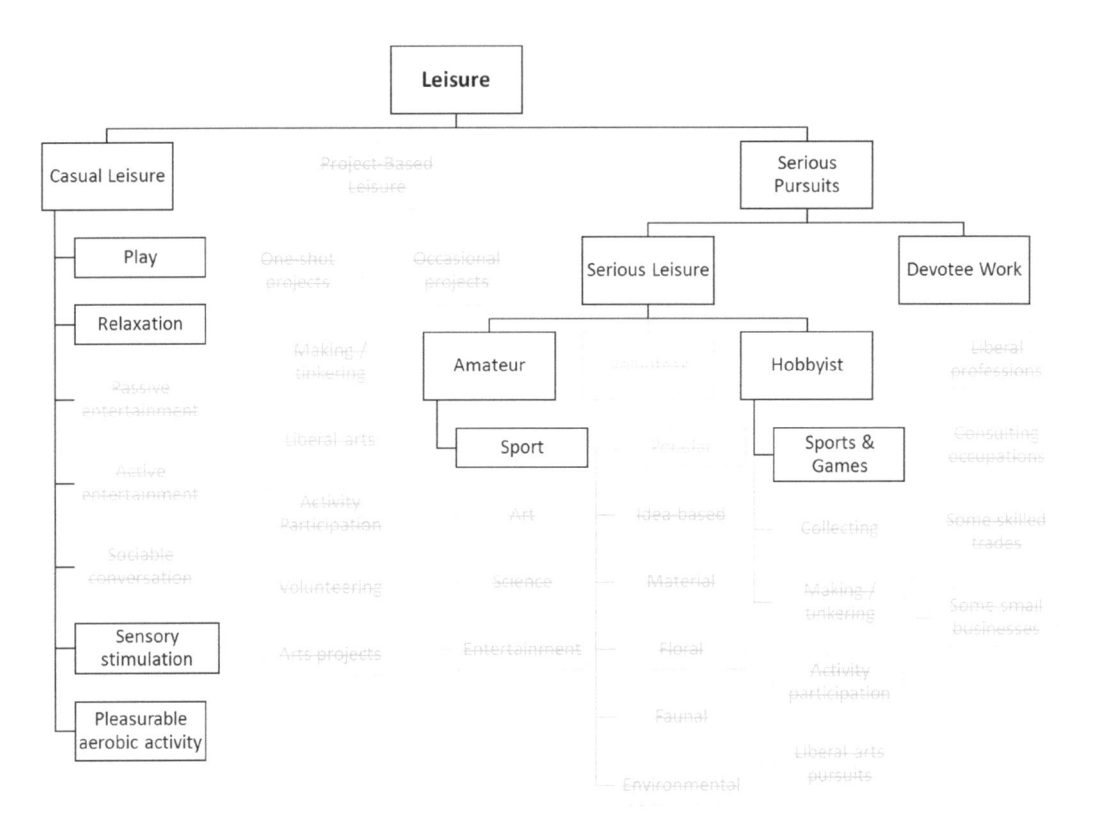

Figure 1. Serious leisure perspective framework (from Hartel, 2013) highlighting categories of human – canine leisure discussed.

as a non-human is precluded from. Hence in 2011 Carmel purchased Bunji to keep Gypsy company while Carmel was at work.

For Carmel being human guardian to two Labrador Retrievers is a central way of life. It means getting up to go walking at 4:00am so that Gypsy and Bunji get their daily walk; standing in the pouring rain in the middle of a cow paddock so they can compete in retrieving trials. Carmel will drive 3000 km for a dryland sled-dog race because Gypsy loves to run, or travel interstate to compete in large flyball competitions because Bunji thrives in the excited atmosphere. Carmel's dogs are highly trained and Carmel's leisure time is dominated by her dogs (leisure) activities. A twist on this story is that Bunji is also Carmel's registered hearing assistance dog and therefore unlike most dogs, also has distinct working times.

Janette, Sookie, Jack and Happi

Our second cross-species team lives a very different life to Carmel and her canine companions. Janette and her three small dogs (Sookie, Jack and Happi) potter around a large house and garden, sharing time on the couch and co-sleeping (Smith et al., 2017). Minimally trained the three dogs do leave the property from time to time but this is very local – to parks within several minutes' walk, to the vet and sometimes to small community events. Janette's leisure life is largely built around time that includes Sookie, Jack and Happi but not focussed on them in the manner that Carmel's leisure involves Gypsy and Bunji.

Janette (also an academic) lives with her human partner and they both work fulltime. The reason for Janette having multiple dogs is the same as for Carmel – so that one dog does not become lonely on its own. Sookie, Jack and Happi are quite different ages however and became part of the pack across time. Jack, a tiny middle aged Papillion came to live with Janette and her first dog when he was elderly and in need of a companion (having lost the canine companion he had grown up with). Sookie the Pomeranian joined Jack and Sputnik when it became obvious that Sputnik would not live much longer and Janette was concerned that Jack would be left on his own. Sookie was already middle aged and is now geriatric herself at 12 years old. Jack and Sookie have cheerfully hung our together for several years and were not impressed when Happi (another Papillion) was introduced into the household early this year as a puppy. As evidenced by physical avoidance, snarling ('maternal snarl' we termed it) and disappointed looks directed at puppy and humans by both dogs.

Results – mapping to the SLP categories of leisure

Carmel, Bunji and Gypsy's leisure time can be seen to fit several sub-categories of both Casual Leisure and Serious Pursuits. Janette's pack maps more specifically to Casual Leisure's.

Casual leisure

'Casual leisure is immediately, intrinsically rewarding; relatively short-lived, pleasurable activity requiring little or no special training to enjoy it.' (Elkington & Stebbins, 2014). Casual leisure can be divided into a number of sub-categories with those identified as pertaining to Gypsy and Bunji being Play, Pleasurable Aerobic Activity, Relaxation, and Sensory Stimulation. Janette's canine pack map to the Casual Leisure's of Relaxation, Passive Entertainment, and Sensory Stimulation most readily, with Play also a dimension for her dogs.

Play

Play can be understood as un-coerced activity undertaken purely for enjoyment (Carr & Young, 2018; Wright, 2016) with aspects of creativity and fulfilment (Sutton-Smith, 2001). The idea that animals engage in a behaviour that would be considered play if they were humans is not new (Beach, 1945) although the purpose and function of play in animals is still not well understood (Miklósi, 2007). In dogs specifically, play has been investigated in intraspecific (dog-dog) and interspecific (dog-human) situations (Sommerville, O'Connor, & Asher, 2017). The opportunity for intraspecies play was a key motivator for Carmel acquiring Bunji. He and Gypsy now frequently engage in spontaneous play both with each other and other dogs when in a larger social grouping of canines. However Gypsy and Bunji don't necessarily play with all the dogs they meet, and the actions of play may vary depending on the dogs they are playing with. The following scenario provides an example of the complexities of play as leisure for these dogs.

Gypsy and Bunji spend a great deal of their time in a group of dogs which form part of their 'obedience club' family. One of these dogs, Jem has always tried to engage Gypsy in play, however has never been successful. The two dogs have travelled, trained and even competed for two years together in sled-dog racing but this familiarity does not necessarily engender a desire in Gypsy to engage with Jem in play. Workmates but not friends perhaps – at least in Gypsy's perspective. Jem will in time turn her attention to Bunji. A move that more often than not will be responded to positively by Bunji with the two often playing on and off for hours engaging in typical behaviours such as play fighting, wrestling, tug-o-war and chasing (Rooney, Bradshaw, & Robinson, 2000). Alternately Gypsy is generally happy to engage in play with a younger male dog called Tex. His canine invitations to play are rarely acted on immediately, but while Gypsy never responds to Jem she will eventually succumb to Tex and the two will have a brief playful encounter. Jem shouldn't feel overly rejected however as Gypsy avoids Bob who heads straight to Bunji anyway, as Bunj is usually receptive to playing with Bob.

In this small snapshot canine choices of *who with* and *when* to engage in frivolous enjoyable activity (play) are demonstrated. While Gypsy is perhaps constrained to work with Jem, in less structured free time she prefers to hang out with Tex. Bunji, with a more laid back and social character is more likely to engage with whoever in this extended network shows interest in playing with him. Individual characters and predilections shape their leisure engagement.

Relaxation

The three dogs who live with Janette do not play much. They will all play ball from time to time, sometimes initiating this by bringing a ball to their human, sometimes initiated by Janette or her partner – but this is not a guarantee of engagement. There does seem to be a game of charging full pelt up and down the long hallway some mornings on the part of Jack and now Happi. The dogs seem to delight in the sense of running and causing a huge level of noise on the wooden floor. The lack of play seems to be at least partly about their spread of ages (given previous experience with two dogs who were close in age), and has also reduced as they have aged, with Happi more likely to initiate requests for play than the older dogs.

Janette's fur-family co-habitate more than actively interacting with each other, although there was certainly an increased sense of camaraderie that emerged between Sookie and Jack on meeting Happi! They will lie either end of the couch, all sleep on the bed at night, and hang out with their humans in the evenings in front of the fire, television, in the study, usually following Janette and her partner around – not necessarily immediately but within a few minutes of them moving rooms they will join them. Their days are filled with snoozing and pottering; heading out through the doggie door to check out the garden from time to time, sniffing garden plants or other objects. Finding the best places to laze in the sun on a good day; or cosy places when cold.

Although they have very active lives, Gypsy and Bunji spend most of their time at home relaxing. Due to their physical size play has always been an 'outside' activity so when indoors both dogs are quick to find a comfortable spot to relax. Even outside one of their favourite places is sitting or sleeping against the shed just as the morning sun creeps over the house, something Carmel has always referred to as 'making an offering to the Sun Gods' (see Appendix Photos).

Pleasurable aerobic activity

Along with a daily walk, Gypsy, Bunji and Carmel engage in a range of informal aerobic activities. Being Labrador Retrievers Gypsy and Bunji both 'love' the water (Mattinson, 2015). This emotive connection is displayed in their actions when they are around water. Put them near water, be it a puddle, creek, river or ocean and both will go into the water without any prompting. In the warm Australian summer it is not uncommon for the trio to head to the beach for a bit of fun. Carmel also brought a kayak so that she was able to enjoy the water with her dogs in combined leisure. Together the trio engage in swimming, kayaking, running, walking – aerobic activity undertaken simply for pleasure.

There is however, a twist on the notion of these things as inherently pleasurable casual leisure. As for some humans, Gypsy' experiences of leisure are impacted by mental health issues (Shor & Shalev, 2016). For Gypsy, some trips to the beach are not as pleasurable as others due to her anxiety. The excitement and play of other dogs, and the approach of some unfamiliar dogs can sometimes be too much for her to the point of triggering panic attacks (Overall, Dunham, & Frank, 2001). She does seem to self-regulate at times, moving away from other dogs to be on her own but sometimes Carmel needs to call her into the water away from other dogs. Here Gypsy's obedience training becomes invaluable as it enables Carmel to care for her by reducing her stress, increasing her (Gypsy's but ultimately also Carmel's) leisure enjoyment.

For Janette, who rides to work each day, dog walking as aerobic activity was not an agenda for seeking canine company. While previous dogs she has lived with have enjoyed going for walks, the current three are intriguing for their responses to what is commonly posited, that 'dogs should be walked every day' (German, 2006; Lim & Rhodes, 2016). Sookie, the eldest, has always had breathing difficulties; this in combination with a belief that she should be carried when outside of home territory (leaping up human legs and sitting down on the end of the leash) walking is not pleasurable aerobic activity for or with her. Jack, the middle aged dog becomes very excited at the sight of the leash, running to the door with his eyes shining, tongue out, leaping up and down. However this excitement tends to wane rapidly. Sometimes just a couple of houses down the street his tail is down and he hunches into himself having been scared by noises and shadows and perhaps other factors. At times he is quivering when it is decided that he needs to be picked up. So again – the aerobic activity of walking does not seem to be inherently leisurely for Jack. Happi, the newest and youngest addition to the pack has not had the chance to go out very often due to the fact that the other two do not like walking. He is showing some interest but not on a regular basis as demonstrated by thwarted attempts to take him for a walk when he skips off to the far reaches of the house when the lead is produced. What these scenarios indicate is that species specific tenets may not match individual beings preferences or experiences of leisure.

Sensory stimulation

Stebbins identifies 'sex, eating, drinking' (Stebbins, 2015a) as examples of the kind of activities that can be undertaken as Casual Leisure – Sensory Stimulation. Touching, through petting, stroking, snuggling up together is an important leisure activity that both Janette and Carmel engage in with their fur families.

Touch is an under-researched topic (Fulkerson, 2014) but the benefits of touch to human health that have been documented are startling. From pre-term babies life expectations increasing dramatically when touched (Barnett, 2005; Field, 1998) to recognising that touch may be the most resilient of our senses as we age (Field, 2014). Human-animal touch has been shown to be reciprocal with physiological benefits for both humans and animals involved in the interaction (King, Buffington, Smith, & Grandin, 2014).

Touch is a key way in which Janette, Sookie, Jack and Happi engage across species. All three dogs will approach Janette for contact in their shared leisure time. Jack and Sookie will both climb onto her chest, lap or shoulder when she sits on the couch. Happi who has not learnt to get up on the couch yet will come up and put his paws on your knee seeking contact. The night time routine involves all three dogs finding their self-allocated space on the bed; cuddling into humans bodies especially when the weather is colder.

Physical closeness and intimacy across species is a core part of Janette and her human partner's leisure and lives together. It is something that is sought across the species divide, although less within the species group with the dogs all seeming to have a respectful distance between each other that if crossed may lead to tensions. Jack and Sookie are amenable to each other being close but not touching – but if they do touch Sookie can get quite cranky – a testiness that Jack is aware of and can be seen seeking to avoid at times. Happi has not learnt to be as careful with regard to this same species touching and in his puppy clumsiness often needs to be reminded of the rules by both Jack and Sookie.

The sensory stimulation of touch and pressure, in combination with auditory soothing are valuable tools for the management of Gypsy's anxiety (King et al., 2014). Gypsy's panic attacks are shorter and less severe when Carmel is able to physically touch and talk to Gypsy. Gypsy will also seek physical contact from Carmel when she is not stressed. It is not uncommon for her to get up off her bed and stand with her chin resting on Carmel for a pat, a position which she has actually fallen asleep in. Both dogs enjoy pats from Carmel, and almost any other human willing to offer. In addition, Bunji will often seek touch from Gypsy. Since puppyhood he has laid with his chin resting on Gypsy's back, and the two are often found sleeping in that position.

Hence across both households there are patterns of cross- and intra-species leisure behaviours that differ on the basis of individual personalities and preferences (Lu & Hu, 2005).

Serious pursuits > serious leisure

It is in the categories of Serious pursuits > Serious leisure > Hobbyist > Sports & Games that the many activities that Carmel and her canine partners pursue can be seen to fit. Gillespie, Leffler, and Lerner (2002) undertook a substantive project focussed on humans who engage in dog sports two decades ago categorising human involvement in these as this SLP category. Our focus is on the *canine* participants of these sports. In the following analysis Stebbins serious leisure involvement scale is also used to unpack the distinctions between Gypsy and Bunjis leisure engagements.

Hobbyist – sports and games

Dog sports can be categorised as 'serious leisure' (as versus casual) due to the fact that they reflect the six features of serious leisure, for both human and dogs. These features are the need to persevere; a 'career' composed of stages of achievement, necessitating effort in order to gain skill and understanding; benefits peculiar to the activity; a social world and identity. While such categorisation may suggest homogeneity across sports and species, Carmel and her fur-family demonstrate a nuanced picture as different dog sports engender varied experiences of leisure across and within species.

While Gypsy and Bunji engage in a number of dog sports they enjoy them differently. Gypsy is high energy and easily excitable (Shabelansky & Dowling-Guyer, 2016). Her drive and motivation for the sport of sled dog racing is completely different to Bunji. While both will show excitement on the start line, Bunji's enthusiasm is short lived and he seems happy just to run 'with Carmel'. Gypsy however will run the entire race at pace and even though tired she is just as enthusiastic and happy at the end as at the start. While the skill required by both dogs and Carmel is identical, their SLP leisure involvement is distinctly different. Gypsy is self-motivated and keen to participate; Bunji might be just as happy to remain a spectator if allowed. Carmel's continued involvement in the sport is driven by Gypsy's interest. Gypsy's canine leisure involvement in sledding can be seen as 'moderate devotee' as she has become competent and in this competence seems to derive great pleasure. Bunji on the other hand is a neophyte. He has developed competence in the activity, but does not show inclination to develop it as a 'career' (Stebbins, 2017).

Flyball is different – with both dogs exhibiting similar leisure responses. Initially Carmel needed a high degree of commitment and motivation to continue the training that has enabled both Gypsy and Bunji to become proficient at the sport. Both dogs seemed to struggle with enthusiasm and focus when introduced to the sport, but now they are at competition standard they require little training to maintain their skills and enjoy the excitement of the sport. There is evidence that the dogs have progressed through the stages of Neophyte, to Participant to perhaps even being moderate Devotees of this sport. Both Gypsy and Bunji seem to have a sense of enjoyment from competing in flyball, and engage with other dogs who also compete creating perhaps some sense of shared canine leisure world and identity. Watching the sport of flyball it is hard to deny that some dogs seem to be competitive in the same way as some humans can be. Dogs will run slowly unless there is a dog in the other lane to run against; some dogs will even look over at the dog they are racing against seemingly to see if they are ahead of them as they cross the finish line. Gypsy and Bunji are part of this competitive culture.

Amateur – sport?

A third sport Gypsy and Bunji engage in is Retrieving. In the casual leisure space of play, as a puppy Gypsy promptly retrieved and returned the first ball Carmel ever threw for her. Untrained she seemed to operate on instinct. However, Bunji never retrieved a ball naturally. Instead he started to do this only when he observed and learnt that when Gypsy returned the ball she was often given a small food reward. Hence it could be argued that for Gypsy, fetching a ball can be the casual leisure of Play, but for Bunji, ball playing is more aptly identified as the Casual leisure -Sensory stimulation of eating.

Playing fetch with a ball however is quite different to retrieving as a dog sport. It is undertaken in a non-domestic environment, with other dogs and humans (creating a social world and identity), and requires higher levels of energy. All of these features can be aligned with the serious leisure of amateur sports. When first introduced to retrieving as a sport, Gypsy and Bunji displayed minimal interest. Retrieving dummies (an object used in retrieving) thrown from short distances was boring and unengaging. However a dummy thrown mechanically 30–50 metres engenders both dogs fixating on the dummy flying through the air to land, running full speed to pick it up and return it to Carmel when given the signal to retrieve. Nowadays it is almost possible to feel the intensity and focus radiating off Bunji (the half hearted ball catcher) and Gypsy in retrieving. For Carmel, the intensity of response displayed by the two in this activity indicates that this form of retrieving is an instinctual need, that as their human guardian she has a responsibility to enable them to access.

Human-animal 'moral leisure'?

A final activity that Carmel and her dogs spend leisure time undertaking is obedience training. Both dogs compete in obedience but neither perform well. However, Carmel perceives obedience training as crucial to responsible human guardianship to ensure she has well behaved and sociable dogs. Obedience training was not initiated with the intention of entering competitions, but as the dogs' skills progressed starting to trial was the only way Carmel felt that she could remain motivated to continue their training. Neither dog seems to enjoy obedience, and to be honest, neither does Carmel. But she feels morally obliged to remain engaged with this activity, even though for all three it feels more like 'work' than leisure.

Discussion

The aim of the explorations presented here has been to delve into the core questions posed at the beginning of this paper. The ability to apply the SLP framework to observable animal experiences in the case studies explored indicates that domestic animal leisure's exist in a manner that can be paralleled to human leisure as posed by our first question. Close observation reveals animals enjoying opportunities for brief, pleasurable activities; developing skills and participating in shared cultures of engagement around sports, and even displaying some behaviours that can be seen to align with 'Devotee work'.

The second question posed was *does human and animal leisure necessarily equate*? We would say that the answer to this is 'no'. While human and animal leisure may at times be shared leisure, human and animal leisure experiences do not always equate. Examples of the misalignment of human and animal; and animal and animal leisure's is shown in these case studies. Carmel struggles to remain engaged in some activities but continues to do so because her dogs, or one of them at least shows enjoyment of an activity. Intra-species, activities that Gypsy enjoys, Bunji tolerates. Jack, Happi and Sookie enjoy many of the same casual leisure activities but their enjoyment hubs around their human companions more than Gypsy and Bunji who enjoy a close same-species relationship and leisure life together, and separate, from their human. Walking is a shared co-species leisure enjoyed by Carmels household. Walking is a single species (human) leisure activity for Janette's troop. Despite animal behaviour mandates – her dogs display individual sentience and agency regarding the notion of dog walks as necessary and 'naturally' enjoyable. Presuming that shared human and animal 'leisure' is necessarily experienced as mutual leisure is disputed by these case studies. But intra-species presumptions of mutual leisure are also challenged.

The third 'question' posed at the beginning of this research was what nuances, distinctions or shades of understanding animal leisure can be teased out? There are two areas of discussion that our research reveals that will be explored here. The first relates to Stebbins SLP categories and the meshing of these with animal specific lives and experiences – how the categories of hobbyist, amateur and devotee worker may be discerned in (domestic canine) animal lives; and secondly the (increasing) application of human parenting and leisure moralising to animal guardianship, or fur-parenting.

Hobbyists or amateurs?

While Gillespie et al. (2002) categorised human engagement in dog sports as the SLP Hobbyist-Sports and Games category, there is evidence that a more nuanced, multi-species understanding of dog sports is possible.

The distinguishing factor between Stebbins Hobbyist Sport and Games and that of Amateur Sport is the lack of a professional parallel for a leisure pursuit. It is possible to argue that canine 'professional' parallels exist for some dog sports. Greyhound racing (a form of animal labour) aligns with Lure coursing. Schutzlund (IPO) is an intense form of obedience training that originated in the training of police and guard dogs. Tracking is core to the 'jobs' of search and rescue or quarantine dogs, while Obedience and Dog dancing have a professional parallel with canine actors. Hence when considering dog sports from the perspective of understanding canine (as versus human) leisure there is a justifiable argument that much of the activity that Bunji and Gypsy (as examples) engage in fits the Amateur Sport category.

In addition when considering animal leisure, instinct, and generationally bred traits need to be included in the frame. Our contention is that breed traits can be added to Stebbins definition of 'Amateur' as activities that have a professional parallel to encompass non-human others who are the product of selective (imposed) generational breeding. For example Retrieving was specifically developed for gundog breeds such as Labrador Retrievers (Bunji and Gypsy's lineage) to test their ability to assist humans in hunting, a role they were specifically bred for (King, Marston, & Bennett, 2012). While most dogs (in western cultures) are now kept as companions not working animals (Asp, Fikse, Nilsson & Strandberg, 2015), the range of breeds reflects generational human management of animal fertility to create animals designed to fulfil human agendas (Notari & Goodwin, 2007; Young & Carr, 2018). Commonly herding, hunting or guarding (King et al., 2012; Svartberg, 2006) but also companionship exemplified by the Papillion breed of Jack and Happi (PetWave, 2015). While retrieving as a sport requires a degree of training in line with the SLP hobbyist – sport and games category – it also meshes with human contrived breed characteristics. Both Gypsy and Bunji exhibit this generationally bred breed homogeneity.

Animal devotee work? Connecting instinct and labour

Building on the contention that impacts of imposed generational breeding should be encompassed in the SLP concept of 'Amateur' we believe that it is possible to extend the concept of Devotee Work to some animal leisure.

An alternate categorisation of Gypsy and Bunji's engagement in Retrieving could be Devotee Work. Both dogs exhibit the kind of intensity and devotion to the activity that in humans could be categorised in this this way. This is the kind of leisure where the line between leisure and labour is indistinguishable (Stebbins, 2015a) and participants feel an intense level of devotion to the activity. Perhaps mirroring the intensity of what can be termed 'instinct' in non-human species. As dependent beings, companion dogs are unable to access such activities independently, hence undertaking 'work' out of hours cannot be demonstrated.

'Instinct' can be seen to connect to a small body of leisure literature considering the possibilities of animal non-leisure or labour needs (Young & Baker, 2018). Bunji and Gypsy's ancestors, as Labrador Retrievers were bred to be co-labourers with humans. Bunji and Gypsy's response to retrieving suggests a now hard wired biological need to undertake tasks they have been bred for. A labour requirement that is now largely redundant. Carmel enables expression of this past labour imperative through supporting her dogs to engage in the dog sport of retrieving. Breed biology may impact and shape animal leisure experiences. Furthermore, as noted previously, there are still some professional employment roles that dogs (http://www.guidehorse.com/; Marks, 2006) may engage in. In undertaking these tasks they are animal labourers, and hence have the potential to be Devotee Workers. As noted previously, Bunji has become Carmels hearing assistance dog, supporting her in her non-leisure and leisure life. Bunji's progression into the role of assistance dog has been built on his obedience training. Akin to some human leisure experiences, engagement and skill development in leisure can lead to employment in these fields (Stebbins, 2015b). In addition when not 'on duty' Bunji will alert Carmel to her phone or other things he has been trained to do as a worker. As with human Devotee Workers his passion for what could be called labour extends into his leisure or off-duty time.

But individual sentient being differences may also indicate non-leisure or Devotee work. Jack, whilst predominately showing characteristics of his breed – gentle and prone to nervousness – has a self appointed role as bird scarer. Any bird that lands on the roof needs to be informed that it should not be there and he barks and leaps until such time as the birds leave. Whilst he displays excitement and some pleasure in this activity (eyes bright, tail up and alert, and a bouncy self-congratulating walk when they leave) his seeming need to undertake this activity in all weather suggests that for him this is a labour undertaking, far more than casual entertainment. Yet the line between leisure and labour in non-human others is particularly difficult to discern as even dogs such as Bunji who labours as an assistance dog, are not formally employed, earning recompense or legally contracted to labour. But our contention is that there is scope to include an Animal Devotee Worker category in the SLP model (as suggested in Figure 2).

Human fur-parents – moralities and motherhoods

Leisure has always been imbued with moral discourses of 'good and proper' uses of free time (Glasser, 1970; Godbey, 1994). More recent explorations have been exploring the complex intersections of gender, health and leisure moralities. Increasingly 'good parenting' is conceptualised as the provision of meaningful, healthful leisure for children (Craig & Mullan, 2012). Expectations of middle class parenting have become more intense (Harrington, 2015). Parenting is expected to provide top up, or prop up to educational opportunities at the same time as further developing

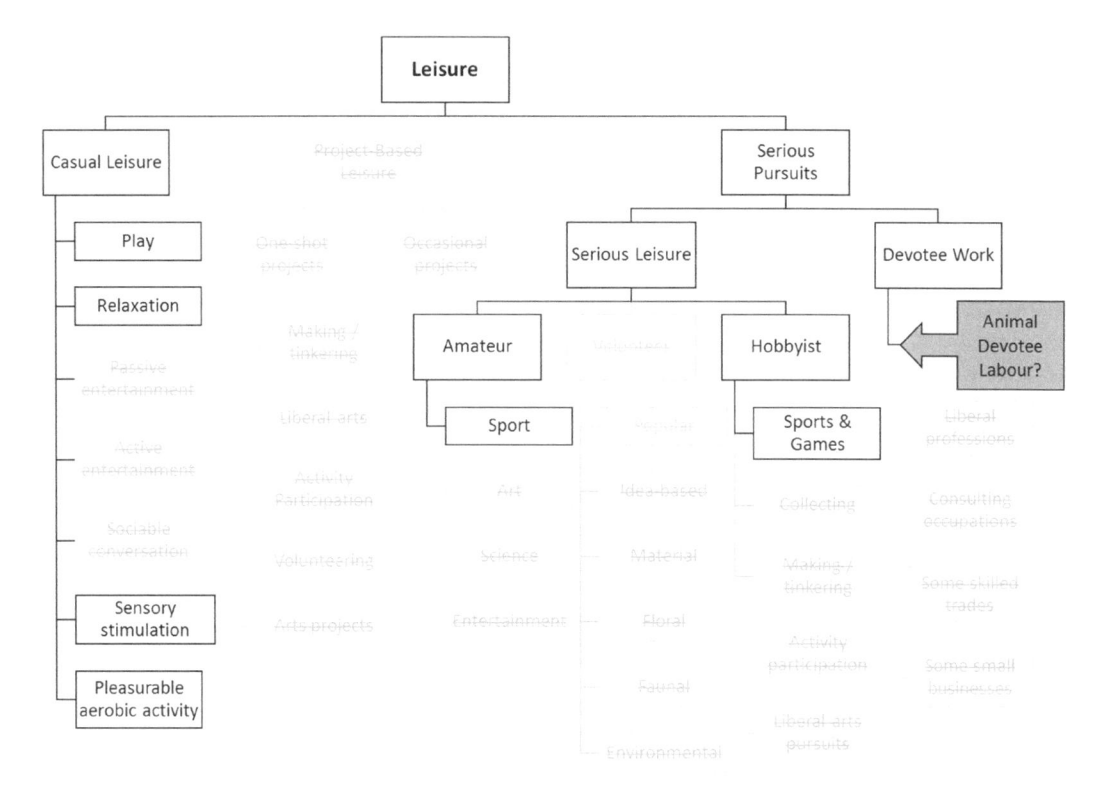

Figure 2. Serious leisure perspective framework (from Hartel, 2013) indicating the addition of an animal devotee worker category.

children's social skills, capacities and abilities (Choi, 2017). Craig and Mullan (2012) explored the notion of 'purposive' leisure with children – leisure time that is pre-planned with aims of enrichment, and nurturing family relationships. Childrens 'free time' has become demonised as time spent in unhealthy activities such as screen time and lack of physical activity (Ming, Merom, Rissel, & Simpson, 2010; Roman-Viñas et al., 2016). Arguably it is possible to see this middle class parenting moralisim being extended to evolving understandings of 'fur families'. Middle class fur-parents are seen to have obligations to provide meaningful leisure for their pets. While our co-species leisure lives are very different, both of us feel and can be seen to be responding to 'fur-parent moralism'.

As has been presented Carmel's leisure time is dominated by her fur babies leisure events. The literature on how parents, especially mothers support their human children's leisure especially sporting engagements (Wheeler & Green, 2012) provides a number of parallels. Carmel as a middle class fur-parent/mother seeks high quality, meaningful and enjoyable leisure for her fur-children. While Carmel does not have the same educational/social concerns linked to future lives as adult humans, when considering Gypsy and Bunji's leisure, her concerns for active, engaging leisure that meets what she perceives to be her dogs needs is clear. She, Bunji and Gypsy engage in Obedience training in order to be good human and canine members of society. Carmel's activities parallel the actions of many mothers of human offspring, with time and energy spent planning routes and time schedules, sourcing and then packing necessary resources into the car, driving to, staying at and then driving home her 'offspring' from leisure destinations. As a good middle class mother she even has some shared leisure activities where she actively engages with her offspring so they have mutually enjoyable leisure (Craig & Mullan, 2012). Carmel's leisure time participation is built around her dogs activities, and she can only undertake independent leisure when their needs are met.

While not overtly or actively engaging with the moralising discourse of fur-parenting in the way that Carmel can be seen to, this moralising discourse impacts on Janette as well. In allowing others to see into her life via auto-ethnography, she has feared she will be accused of lack of care for her three fur-companions. She has constructed arguments for why she chooses small dogs, and why they don't get out much ('they don't need extra exercise living in such a big space and they have each other'). Taking Happi to puppy training this year (as a good-puppy mother) she was encouraged to ensure that her puppy ate correctly, was mentally stimulated, and taught the basics of polite dog behaviour. In combination with recognising the power imbalance inherent in the pet-human relationship (ultimately we, ever so loving human guardians control our fur-friends choices) the good middle class fur-parent narrative impacts on her thinking and psyche perhaps with no less impact than for Carmel.

Societal discourses have the power to shape our ways of seeing, understanding and applying values in our everyday lives (Fairclough, 2003). The content of these discourses may not be inherently harmful, and may indeed offer positive, pleasing and enjoyable outcomes for some. But they may simultaneously be intensely stressful as parents feel conflicted about meeting require-ments of good parenting, and homogenising discourses overlook individual leisure interests, imposing moralised ideals of opportunities onto sentient others – animal and human.

Conclusion

Companion animal leisure is enmeshed in human guardian lives, choices and leisure. Yet as shown here presumptions as to animal leisure cannot be made simply on the basis of breed, instinct, moralism or co- and same species sharing of activities. Individual sentience – preferences, person-alities and choices shape what can be observed of animal experiences of the human-contrived concept of 'leisure'. While the research presented here focusses on companion canines there is potential for many if not all our analyses to be transferred to considerations of the leisure of other, particularly domestic species. The SLP framework offers a means of shaping these explorations, and furthering our understandings of multi-species leisure in the shared lives of humans and animals. Understandings that can enhance leisure experiences across and within species.

Acknowledgments

The authors would like to acknowledge the animal-others who are the focus of this work. They receive no payment, or status from this, but our lives are enriched by their presence and engagement in ours.

Disclosure statement

No potential conflict of interest was reported by the authors.

ORCID

Carmel Nottle (iD) http://orcid.org/0000-0001-8649-3552
Janette Young (iD) http://orcid.org/0000-0002-2284-3485

References

Animal Medicines Australia. (2016). Pet ownership in Australia 2016. Retrieved from https://animalmedicinesaus tralia.org.au/wp-content/uploads/2016/11/AMA_Pet-Ownership-in-Australia-2016-Report_sml.pdf

Aps, H. E., Fiske, W. F., Nilsson, K., & Strandberg, E. (2015). Breed differences in everyday behaviour of dogs. *Applied Animal Behaviour Science, 169*, 69–77.

Barnett, L. (2005). Keep in touch: The importance of touch in child development. *International Journal of Infant Observation and Its Applications, 8*(2), 115–123.

Beach, A. (1945). Current concepts of play in animals. *The American Naturalist, 79*(785), 523–541.

Carr, N., & Young, J. (2018). Conclusion: Charting a way forward. In N. Carr & J. Young (Eds.), *Wild animals and leisure: Rights and wellbeing* (pp. 225–233). Abingdon, UK: Routledge.

Choi, K. W. (2017). Habitus, affordances, and family leisure: Cultural reproduction through children's leisure activities. *Ethnography, 18*(4), 427–449.

Cochrane, A. (2016). Labour rights for animals. In R. Garner & S. O'Sullivan (Eds.), *The political turn in animal ethics* (pp. 15–32). London, UK: Rowman and Littlefield.

Craig, L., & Mullan, K. (2012). Shared parent–Child leisure time in four countries. *Leisure Studies, 31*(2), 211–229.

Danby, P. (2018). Post-humanistic insight into human-equine interactions and wellbeing within leisure and tourism. In J. Young & N. Carr (Eds.), *Domestic animals, humans and leisure* (pp. 165–189). Abingdon: Routledge.

Dawkins, M. S. (2006). Through animal eyes: What behaviour tells us. *Applied Animal Behaviour Science, 100*, 4–10.

Elkington, S. D., & Stebbins, R. A. (2014). *The serious leisure perspective: An introduction.* Abingdon, UK: Routledge.

Fairclough, N. (2003). *Analysing discourse: Textual analysis for social research.* London, UK: Routledge.

Field, T. (1998). Touch therapy effects on development. *International Journal of Behavioral Development, 22*(4), 779–797.

Field, T. (2014). *Touch.* Cambridge, MA: MIT Press.

Franklin, A. (2015). Miffy and me: Developing an auto-ethnographic approach to the study of companion animals and human loneliness. *Animal Studies Journal, 4*(2), 78–115.

Fulkerson, M. (2014). *The first sense: A philosophical study of human touch.* Cambridge, MA: MIT Press.

Geertz, C. (1973). *The interpretation of cultures: Selected essays.* New York: Basic Books.

German, A. J. (2006). The growing problem of obesity in dogs and cats. *The Journal of Nutrition, 136*(7), 1940S–1946S.

Gillespie, D. L., Leffler, A., & Lerner, E. (2002). If it weren't for my hobby, I'd have a life: Dog sports, serious leisure, and boundary negotiations. *Leisure Studies, 21*(3–4), 285–304.

Glasser, R. (1970). *Leisure: Penalty or prize?* London, UK: Macmillan.

Gobo, G. (2008). *Doing Ethnography.* London: SAGE Publications.

Godbey, G. (1994). *Leisure in your life: An exploration.* Pennsylvania, USA: Venture Publishing.

Graham, T. M., & Glover, T. D. (2014). On the fence: Dog parks in the (un)leashing of community and social capital. *Leisure Sciences, 36*(3), 217–234.

Harrington, M. (2015). Practices and meaning of purposive family leisure among working- and middle-class families. *Leisure Studies, 34*(4), 471–486.

Hartel, J. (2013). Diagrams of the serious leisure perspective. Retrieved from https://www.seriousleisure.net/slp-diagrams.html

Hultsman, W. Z. (2012). Couple involvement in serious leisure: Examining participation in dog agility. *Leisure Studies, 31*, 231–253.

King, C., Buffington, L., Smith, T. J., & Grandin, T. (2014). The effect of a pressure wrap (ThunderShirt®) on heart rate and behaviour in canines diagnosed with anxiety disorder. *Journal of Veterinary Behaviour, 9*, 215–221.

King, T., Marston, L. C., & Bennett, P. C. (2012). Breeding dogs for beauty *and* behaviour: Why scientists need to do more to develop valid and reliable behaviour assessments for dogs kept as companions. *Applied Animal Behaviour Science, 132*, 1–12.

Lim, C., & Rhodes, R. E. (2016). Sizing up physical activity: The relationship between dog characteristics, dog owners' motivations, and dog walking. *Psychology of Sport and Exercise, 24*, 65–71.

Lu, L., & Hu, C.-H. (2005). Personality, leisure experiences and happiness. *Journal of Happiness Studies, 6*, 325–342.

Madden, R. (2014). Animals and the limits of ethnography. *Anthrozoos, 27*(2), 279–293.

Marks, D. (2006, January 14). Monkey helpers lend a 'helping hand' [CBS News]. Retrieved from https://web. archive.org/web/20060927041542/http://www.klas-tv.com/Global/story.asp?S=4361694.

Mattinson, P. (2015). *The labrador handbook. Your definitive guide to care and training.* Great Britain: Ebury Publishing.

Miklósi, A. (2007). *Dog behaviour, evolution and cognition.* Oxford, UK: Oxford University Press.

Ming, W. L., Merom, D., Rissel, C., & Simpson, J. (2010). Weight status, modes of travel to school and screen time: A cross-sectional survey of children aged 10–13 years in Sydney. *Health Promotion Journal of Australia, 21*(1), 57–63.

Notari, L., & Goodwin, D. (2007). A survey of behavioural characteristics of pure-bred dogs in Italy. *Applied Animal Behaviour Science, 103*, 118–130.

Overall, K. L., Dunham, A. E., & Frank, D. (2001). Frequency of nonspecific clinical signs in dogs with separation anxiety, thunderstorm phobia, and noise phobia, alone or in combination. *Journal of the American Veterinary Medical Association, 219*(4), 467–473.

PetWave. (2015). Papillon – History and health. Retrieved from https://www.petwave.com/Dogs/Breeds/Papillon/ Overview.aspx

Roman-Viñas, B., Chaput, J., Katzmarzyk, P. T., Fogelholm, M., Lambert, E. V., Maher, C., … Tremplay, M. S. (2016). Proportion of children meeting recommendations for 24-hour movement guidelines and associations with adiposity in a 12-country study. *International Journal of Behavioural Nutrition and Physical Activity, 13* (123), 1–10.

Rooney, N. J., Bradshaw, J. W. S., & Robinson, I. H. (2000). A comparison of dog-dog and dog-human play behaviour. *Applied Animal Behaviour Science, 66*, 235–248.

Saunders, D. M., & Turner, D. E. (1987). Gambling and leisure: The case of racing. *Leisure Studies, 6*(3), 281–299.

Scott, D., & McMahan, K. K. (2017). Hard-core leisure: A conceptualization. *Leisure Sciences, 39*(6), 569–574.

Shabelansky, A., & Dowling-Guyer, S. (2016). Characteristic of excitable dog behaviour based on owners' report from a self-selected study. *Animals (Basel), 6*(3), 22.

Shor, R., & Shalev, A. (2016). Barriers to involvement in physical activities of persons with mental illness. *Health Promotion International, 31*(1), 116–123.

Siderelis, C. (2001). Incidental trips and aquarium benefits. *Leisure Sciences, 23*(3), 193–199.

Smith, B., Hazelton, P., Thompson, K., Trigg, J., Etherton, H., & Blunden, S. (2017). A multispecies approach to co-sleeping integrating human-animal co-sleeping practices into our understanding of human sleep. *Human Nature-An Interdisciplinary Biosocial Perspective, 28*(3), 255–273.

Sommerville, R., O'Connor, E. A., & Asher, L. (2017). Why do dogs play? Function and welfare implications of play in the domestic dog. *Applied Animal Behaviour Science, 197*, 1–8.

Stebbins, R. (2017). *Leisure's legacy: Challenging the common sense view of free time.* Cham, Switzerland: Springer Nature.

Stebbins, R. A. (2013). The serious leisure perspective. Retrieved from https://www.seriousleisure.net/

Stebbins, R. A. (2015a). *Leisure and the motive to volunteer: Theories of serious, casual, and project-based leisure.* Basingstoke, UK: Palgrave Macmillan.

Stebbins, R. A. (2015b). *Serious leisure: A perspective for our time.* Piscataway, NJ: Transaction Publishers.

Sutton-Smith, B. (2001). *The ambiguity of play.* Cambridge, MA: Harvard University Press.

Svartberg, K. (2006). Breed-typical behaviour in dogs – Historical remnants or recent constructs. *Applied Animal Behaviour Science, 96*, 293–313.

Veal, A. (2017). The serious leisure perspective and the experience of leisure. *Leisure Sciences, 39*(3), 205–223.

Wheeler, S., & Green, K. (2012). Parenting in relation to children's sports participation: Generational changes and potential implications. *Leisure Studies, 33*(3), 1–18.

Wright, J. T. (2016). The interrelationship of leisure and play: Play as leisure, leisure as play. *American Journal of Play, 9*(1), 89–91.

Yang, H., Kim, J., & Heo, J. (2018). Serious leisure profiles and well-being of older Korean adults. *Leisure Studies,* Advanced online publication. doi: 10.1080/02614367.2018.1499797.

Young, J., & Carr, N. (2018). Domestic animals' leisure, rights, wellbeing: Nuancing 'domestic' asymmetries and into the future. In J. Young & N. Carr (Eds.), *Domestic animals, humans and leisure: Rights, welfare and wellbeing* (pp. 209–222). Abingdon: Routledge.

Young, J., & Baker, A. (2018). From labour to leisure: The relocation of animals in a modern western society. In J. Young & N. Carr (Eds.), *Domestic animals, humans and leisure: Rights, welfare and wellbeing* (pp. 209–222). Abingdon: Routledge.

Young, J., & Nottle, C. (2017). *I wasn't disabled until I got an assistance dog: Human pathologisation meets animal reconfiguration.* Paper presented at Australasian Animal Studies Association Conference 2017 (AASA), July 3–5, Adelaide, South Australia.

Tuesdays with Worry: appreciating nature with a dog at the end of life

Justin Harmon

ABSTRACT

Dogs have been called 'social lubricants' for their uncanny ability to help people with serious illnesses and trauma find brief moments of catharsis, create meaningful relationships when there are none and connect with healthcare providers and other support team members for those with serious illnesses. In this paper, the therapeutic qualities of human–dog interactions will be demonstrated by focusing on one woman's terminal cancer diagnosis while in end-of-life care through her shared leisure experiences with a dog in a natural environment. This auto/ethnography sought to exhibit the simple importance of a dog to the meaning-making process for someone coming to terms with their mortality. Complementary therapies and support are essential to those receiving medicalised care, but for those beyond the treatment stage, the important and necessary support is to be found in the relationships and activities that are of the most significance. As is demonstrated, dogs can provide this necessary kind of therapeutic support that may not be found elsewhere.

Diane[1] didn't say much while I was speaking to the Stage IV cancer support group. In fact, the only sound I heard from her was the low rumble from her oxygen supply. She listened intently, though, and when I mentioned that my dog (Worry) came out on all the hikes with me, her eyebrows raised and she became more attentive; it was about then that she shut off the valve to her air supply so she could listen better. I gave everyone in attendance a flier for the hiking program, but I didn't expect many to be able to participate. As I was getting ready to leave the building, I heard a faint beeping sound slowly coming towards me. It was Diane's oxygen concentrator making the noise, and when I turned around she said to me, 'I sure wish I could go on a hike with you, but I haven't got the energy anymore. But I really like being in nature and I love dogs. If you'd ever be interested in going to a park sometime with Worry, I would be delighted.' – From the author's journal

Introduction

Canines have been faithful human companions for an estimated 15,000 years (Yong, 2015), and their connection to humans has only grown more intimate, with many referring to their beloved dog as a member of their family (Carr, 2014). And while dogs have been employed in numerous working capacities over their long history, it is only in the last half century that dogs have become increasingly embraced for their therapeutic potential in helping people with a wide array of health needs (Johnson, Meadows, Haubner, & Sevedge, 2008).

Dogs have been called 'social lubricants' (Moody, Maps, & O'Rourke, 2002) for their uncanny ability to help people with serious illnesses and trauma find brief moments of catharsis (Browder,

2009), create meaningful relationships when there are none (Fox & Gee, 2017) and connect with healthcare providers (White et al., 2015). In this paper, the therapeutic qualities of human-dog interactions in natural environments will be demonstrated by focusing on one woman's terminal cancer diagnosis while in end-of-life care (Engleman, 2013).

Kleiber, Hutchinson, and Williams (2002) highlighted the potential of leisure to be used as a coping mechanism for those with serious and terminal illnesses because it can help buffer against the negative emotions that often accompany traumas. Janke and Jones (2016) concurred, suggesting that leisure can help people transcend the loss of *another* person. Here it will be displayed that leisure, in the form of meaningful time spent with a dog in nature, can be of great significance to the person who is navigating the meaning-making process while coming to terms with their mortality. As will be displayed, what started out as me volunteering to help someone with terminal cancer get out into nature turned into a new friendship—between the patient and my dog.

Literature review

Cancer and the end-of-life

It is common for people near the end of their life to surrender roles and reduce or eliminate their participation in once enjoyable activities and relationships (Warne & Hoppes, 2009). Once a cancer diagnosis becomes terminal, the patient may feel as if they are a burden to loved ones and lose a sense of dignity about themselves (Chochinov et al., 2005). Because of this, the main purpose of palliative care is to aid in maintaining the highest quality of life (QOL) possible for the patient during their remaining time, typically defined by the patient's preferences (Fegg et al., 2010). Often these QOL markers are related to meaningful leisure activities that have held significance over the course of one's life (Hendricks, 2012).

Leisure has been demonstrated as a coping mechanism for traumatic life events and is believed to be most effective when it is personally relevant (Hutchinson, Loy, Kleiber, & Dattilo, 2003; Kleiber et al., 2002). The opportunity to participate in meaningful leisure, especially near the end of life, can help the patient find meaning in the face of their illness (Krause, 2009). If patients are still able to find purpose when they are dying through some form of leisure, it can improve coping mechanisms as well as alleviate aspects of physical symptoms (Tomás-Sábado et al., 2015). Often this new meaning involves letting go of the 'unattained' goals and focusing on new ones that are more realistic based on the limitations brought on by their health status (Martin & Kleiber, 2005). This requires a shift in focus and a positive reorientation to make the most of what is within reach; this can often be a highly distressing process (Chochinov et al., 2009). However, the creation of a holistic approach to care can ease that sense of distress greatly (Chochinov et al., 2005). For those who found meaning through the outdoors pre-diagnosis of cancer, there still remains significant benefit for them to find with continued exposure to natural environments post-diagnosis of cancer (Song, Ikei, & Miyazaki, 2016).

Fegg et al. (2010) found that people in palliative care find high levels of personal meaning through exposure to nature and interactions with animals, though people in this stage often have less opportunity or ability to fulfil those pleasures. It has been documented that immersion in nature is closely associated with meaning making for those nearing the end of life (Nakau et al., 2013), suggesting that exposure to natural environments can be a necessary form of therapeutic support for some. Related, people who have experienced a 'close brush with death,' such as a terminal diagnosis of cancer, report high levels of appreciation for being in nature to reorient and find some semblance of peace in their life (Martin & Kleiber, 2005). Those who receive benefit from exposure to natural environments, and are also comforted by the company of dogs, may find the synthesis of therapeutic natural landscapes with a dog to be of significant value to their state of being, regardless of their health prognosis (White et al., 2015).

Human-dog bonds

Numerous studies support the claim that spending time with a dog leads to a sense of positive wellbeing in one's life (Charles, 2014, 2016; Duvall Antonacopoulos & Pychyl, 2014; Haraway, 2003; Smith, Treharne, & Tumility, 2017). Charles (2016) and Haraway (2003) have established the need to explore the dynamic of the dog-human bond more intimately, suggesting that we live in a post-humanist society and should embrace the reality that for many people, the dog(s) in their lives are as loved and essential to their happiness as any human could be. Haraway (2003) put forth the notion that, in our ever-evolving relationship with our beloved dog companions, we need to recognise them as adults, albeit of another species, who are capable of having a real relationship with us, and not simply reduce them through infantilising or overtly anthropomorphising behaviours or interactions. Because dogs are capable of reciprocating the feelings and mood of their guardians, their companionship and presence in our lives can be of paramount importance (Charles, 2014). With dogs' innate ability to sense human emotion and improve mood, they can also be important members of a support team in the face of harrowing life circumstances like serious illness (Crossman, Kazdin, & Knudson, 2015).

The therapeutic properties of dogs

Evidence suggests that dogs have 'agency' (Carter & Charles, 2018; Irvine, 2004), thus indicating their potential to make real, lasting connections with humans when feelings of affection are reciprocated (Fox & Gee, 2017). Because of this innate ability to connect with people, dogs often make for ideal 'therapists' in helping people cope with illnesses and traumas (Creagan, Bauer, Thomley, & Borg, 2015; Marcus, 2012; Moody et al., 2002; White et al., 2015). Specifically for people with advanced stage cancer, dogs have been shown to provide comfort and lower stress for those receiving chemotherapy (Chubak et al., 2017; Urbanski & Lazenby, 2012). Equally notable, it has been shown that pain severity decreases after dog visits to outpatient clinics as well (Marcus et al., 2012). All of this evidence suggests that the benefits of human-canine interactions are significant. While the preceding studies are indicative of animal-assisted therapies (AAT), goal-directed interventions where a dog is part of the treatment process (Horowitz, 2010), dogs can be beneficial in less structured and non-medicalised environments as well.

Animal-assisted activities (AAA) include casual activities that involve dog visits, but are not focused on the treatment process (Horowitz, 2010), though may still have significant benefit to the patient or client. For those who have difficulty in speaking about their illness or fragility, dogs can serve as mediators for facilitating the expression of emotions and concerns (White et al., 2015). Dogs can be a 'catalyst' to start conversations that are difficult to have for people with cancer (Browder, 2009), thus improving the relationship between care provider and patient (Johnson et al., 2008). Not all health needs are attended to exclusively in medicalised settings or venues, therefore there remains a need for coping resources outside of hospitals and clinics, too, especially in the places where patients engage in a preferred leisure activity (Harmon, 2018). Building on the unique relationships that are often established between humans and dogs (Kulick, 2017), this manuscript seeks to illustrate how one person with terminal cancer built a friendship with a dog while spending time in nature.

Methods

Background of study

I started a hiking programme in late 2016 for those affected by cancer. The hiking programme is not intended to focus on the illness experience, but instead the 'normalisation' process of life with cancer or life after cancer. The group hikes year-round, every Wednesday and Saturday. Midweek hikes are shorter in duration and are designed to be at a slower pace. These hikes are geared towards those who may have stamina, balance or lower levels of physical ability, as well as those whom are no longer working. Saturday hikes are longer and at a faster pace and are accessible to those with higher levels of physical ability. The majority of participants are referred by oncologists and clinical social workers from the nearby cancer centre. Additionally, I also speak at numerous events at the cancer centre, whether it be type-of-cancer specific groups (i.e., breast, prostate), or the post-treatment programmes whose intent is to promote active and healthy lifestyles either after diagnosis or after patients have gone into remission. Therefore, other participants come from these forums as well.

Auto/ethnographic approach

Autoethnography draws heavily from the personalised accounts of one's phenomenological experience in order to craft an understanding of what has taken place (Denzin, 2006). Often autoethnography has deeply personal and emotional components involving introspection that turn the narrative into the method (Ellis, 2004). However, because this essay was crafted on the story of Diane's interactions with my dog, there are elements of duoethnography here as well (Breault, 2016). Duoethnography is traditionally understood as two researchers working together to build a narrative. In this instance, however, my auto/ethnographic understanding of what took place was a reflection of Diane and Worry's 'conversations' to construct the story I tell here (Taylor, 2007). Norris and Sawyer (2012) said that in duoethnography, 'the journey is mutual and reciprocal' (p. 13), so in that sense, Diane's story of her illness and mortality, as told to my dog, indicates a cooperative sharing of the meaning and embodiment between myself and Diane, and certainly between she and Worry as well (Charles, 2016). This was further substantiated in the conversations between Diane and me on the numerous car rides home. This is also where member checking took place and trustworthiness was established (Tracy, 2013), as I would delicately ask for clarifications about what I heard and how I interpreted it, but also get Diane's teasing of me for tiptoeing around the issues related to her health and mortality. Because of this, the manuscript is in many ways co-constructed (Ellis, 1998).

Perhaps more accurately though, this is neither autoethnography or duoethnography, but auto/ethnography, a hybrid or derivative of narrative ethnography (Gubrium & Holstein, 2008). Gubrium and Holstein said of narrative ethnography that it is the 'social dimensions of narratives... that take us outside of stories and their veridical relationships to storytellers and experience' (p. 250). I initially did not approach these outings with Diane and Worry as 'research'—that is until I saw the relationship and its healing properties that unfolded. And even as I started to think about my roles as chaperone and voyeur for these two friends, I wondered just what sort of worthwhile scholarship might come from it. Ellis (2004) stated that anyone undertaking a form of autoethnographic research was 'going into the woods without a compass' and would need to take time to find their bearings and get 'the lay of the land' (p. 120). This was true metaphorically, and in some ways, literally as well.

The rationale, then, for embracing the auto/ethnographic approach is an attempt to bridge my perspective and understanding of what transpired as a researcher with Diane's perspective of coming to terms with her mortality. However, there is clearly another very important individual involved in this story: my dog, Worry. Hamilton and Taylor (2017) warned against overvaluing the human in human-animal interactions, and this is something I have tried to respect in the representation to follow. Worry and Diane are the central figures in this essay, and it goes without saying that *their* story could not be told without both as they are central figures in what transpired.

Other autoethnographic forays into interactions with those near death have chronicled the conflicting accounts of making sense of someone else's mortality (*cf.* Warne & Hoppes, 2009). As an auto/ethnographer, it is my duty to provide enough detail to enable the readers to determine the transferability of the phenomena to their lives (Ellis, 1998). Therefore, this is not an attempt to tell *Diane's* story, but to chronicle the importance of how someone facing the end of their life derived great benefit from exposure to natural environments and time spent with a friendly dog while coming to terms with mortality.

Tuesdays with Worry

I picked up Diane the following Tuesday to go out to the marina and sit at a picnic table that looks over the water and a forested area popular for its trails. We started to get acquainted on the ride over and much of our first discussions involved the basic 'getting to know you' back-and-forth common to those newly acquainted. I could tell she was very interested in Worry, and while my pup is very friendly, sometimes she can be more interested in what's going on outdoors than other people. Usually when we come to the marina it is for a hike; I think Worry was a little bummed we weren't off in the woods this Tuesday. While Diane slowly and softly petted her, Worry looked off into the forest, though she seemed content with the attention. – From the author's journal

Meeting a friend

I picked up Diane on most Tuesdays for about six months. The first time I spoke with her, in the hallway after the Stage IV cancer meeting, she mentioned that she had recently been given 'a year of good living left' as she put it. A few months after that statement, I was curious if that diagnosis held true, or if it had been updated—for better or worse. Diane (71 years old) was on her second cancer diagnosis. The first was breast cancer in her early 50s; the second was lung cancer that metastasised to her bones at 68 years old.

When I picked her up the very first time, I asked her about her past experiences 'outdoors' and her love of dogs. She was brought up in rural North Carolina as a 'tomboy,' always playing outside with her brothers and cousins on the family farm. She credits her upbringing, and the outdoors, for making her aware of the importance of taking care of the environment, though as she put it, sometimes she did not always practice what she preached. She felt the older she got the more she respected wildlife and wildlands, and felt especially strongly about society's responsibility to take care of its natural resources. While she had not spent much time immersed in the wilderness in years, she felt it always 'called' to her. Diane found it to have a calming influence and that was something she hoped to get out of our outings.

Dogs, though, were *very* important to her. They had been part of her family ever since she was a child, and like nature, the older she got, the more she appreciated them. Her last dog had died shortly after her second cancer diagnosis, and that made the illness even more difficult for her. Diane's family had encouraged her to get another dog; they thought it would help her cope with her new diagnosis, but she was concerned the dog would outlive her—or worse, that she simply would not be able to take care of it. She did not want to build a bond with another dog only to have to leave it to someone else. Because of this, she said her biggest inspiration to come out was to get to hang out with a dog on a regular basis and not feel obligated to take care of it if she were ever unable to. I imagine she did not foresee such a close bond coming from their interactions.

Down by the water

After a few Tuesdays in a row (we always went out on Tuesdays), I noticed that Diane was building an endearing rapport with Worry. I consciously started to remove myself from the dialog and let them interact more, with me slowly scooting to the other side of the picnic table as I looked out over the lake. Worry had also grown less 'worried' about not being off in the woods on a hike; she seemed to enjoy this special regularly scheduled attention from Diane. And while Diane and I hadn't talked too much to one another about her illness and fate to that point, she had no problems speaking with Worry about it. And Worry, seemingly, would just listen, often with her head on Diane's lap, getting lazily stroked, lulled to sleep. – From the author's journal

We settled on visits to the marina for a few reasons, but mostly because it had a clear view of the water and abutting forest, as well as decent shade and seating. Diane did not like sitting on the hardwood planks of the picnic table, they put pressure on her bones and joints and made her ache, so on our second visit she brought a blanket with her, though that did not appear to offer much comfort. By the third visit I had dug out an old collapsible nylon camping chair that set low to the ground. She found that to be the most comfortable, but it also made it harder for her to get up. Because of this I had to learn a gentle way of helping her up, something akin to picking a drunk up off their butt. It was awkward at first, but after several weeks it became second nature. She no longer grunted because I was no longer unintentionally, and often somewhat carelessly, hip-checking her as I tried to get leverage. Worry always appeared to find this act amusing. A half-cocked head and a slowly wagging tail indicated that she assumed I did not know what I was doing. She was correct.

In the early weeks I sat next to Diane on the uncomfortable picnic bench, likely just as uncomfortable as she had been because of my boney butt, though I tried not to let on this was so. My goal was for her to be as comfortable as possible, and I wanted each experience to be enjoyable. At the beginning, though, I had no clue what would be enjoyable to her and what would not. I am not typically one to sit idly at a park, nor is Worry, so we both had to learn to find comfort in our new activity. This was a well-needed lesson in patience for both of us.

It did not take long for the routine to become comfortable for us, and since we were never there for more than 30 minutes, and rarely ever that long, I adjusted easily to the new weekly routine. Worry had also accepted that Tuesdays were for relaxing, and with her comfy dog bed laid next to Diane's camping chair, they both appeared quite content. It was me who was the last to acquiesce. This was in part due to physical discomfort, and in part due to my fidgety nature, but more than anything it was because I always felt as if I had to talk to Diane. She was too polite to tell me there was no need for my constant chatter; eventually I learned that she simply wanted to look over the water to the forest beyond it and pet my dog. And as the Tuesdays rolled by, for those short 20 minutes or so, I had to wonder if Worry was mine or hers during those moments. Worry apparently did not feel the need for my constant chatter or attention either.

Confiding in Worry

After a couple of Tuesdays in a row where inclement weather and Diane's health prevented us from going out, she got in my car far our next outing and Worry almost jumped into her lap. This was, of course, after she had started whining at the sight of Diane slowly ambling down her walkway with oxygen tank and cane in tow. I took this as a good sign. Diane, barely acknowledging me said, 'There's my sweet girl! Oh, how I've missed you!' Worry laid down in the back and Diane chitchatted with me on our drive to the marina. She confided in me that these brief outings were the premier highlight of her week, and she really missed not being able to get out the two weeks prior. She even indicated that she felt worse for having not had our routine. Diane said, 'You know, weather and grogginess be damned. I don't want to miss another Tuesday with my girl!' – From the author's journal

Eventually I took the hint—as loving as it was—that Diane was there at the marina solely to be with Worry. She and I talked on the car ride to and from, and enjoyed one another's company, but I finally deduced that she had created a sacred space where only one other being was needed— my dog. And that was fine. More than fine. By the fourth or fifth week Diane started having playful conversations with Worry, and I found them endearing, though at first none were very introspective or impactful, just the typical 'Who's a good girl?' talk we all have with our pups. Eventually, though, she started to confide in Worry about her memories, fears, and the things she wished she could do but never would. I do not believe these revelations were meant for my ears, even though I was mere feet away; there was never a glance my way nor did Diane ever directly address me during these moments. Many times the things she would say to Worry were very heartfelt and emotional; I constantly found myself with blurry eyes, barely capable of holding back the tears. I never saw Diane cry, though; by this point she had accepted her fate, even if she did not know exactly how much longer she had.

The first 'conversation' she had with Worry really stuck in mind, likely because the statement was so powerful and I had not expected it. Diane said, 'Worry, I'm not ready to die, but I guess it's not my choice, is it?' At that moment I felt like I should say something, but knew that was not what Diane would have intended for me to do, nor would I be able to offer anything that could assuage those concerns of hers. The only response she was looking for was a comforting glance from Worry, which she received, and then Worry closed her eyes and let out a deep sigh of comfort.

The reversal of seasons

By late in the fourth month of Tuesdays with Worry, just as the winter nip was starting to retreat from the air, and the buds began to peak their heads out of tree limbs everywhere, Diane finally started to show a downtick in her health. She had always been frail, and to look at her from day one you would have known she was dealt a heavy hand, but now her colour was dissipating and her movements, while always slow, were now quite measured and burdensome. It was as if she felt one wrong move would be her last. This may have been so since the cancer had penetrated much of her skeletal structure in her hips and lower back—she had become the china shop and the world the bull. She also had graduated from cane to walker, though she had not lost her sense of humour: in place of the traditional tennis ball on each leg she had indecipherable animal tails. I never asked, but I assumed they were dog tails.

Nevertheless, she was just as adamant about going to our spot (or more accurately, *their* spot) every Tuesday. While her health had gotten worse, her constitution seemed stronger: if I was available, she was going out. No discussion to be had. She stuck to her guns about her earlier claim from a few months earlier, there would be no Tuesdays missed on her account.

And now that the weather was changing for the better, and spring was coming into full bloom, her connection to the environment became that much more apparent. Her connection to Worry had always been there, but it was not until roughly four months in that her love of the outdoors was visually reawakened, almost as if it also came out of hibernation. She began to bring a few books with her, stuffed in an old, worn-out, reusable cloth grocery bag, one about birds of the region and the other about trees in the area. They looked well-worn, dog-eared and highlighted, creased with broken spines. And as her health grew worse, her interactions with me grew more prevalent and intentional; at least in the car. The marina still involved little interaction between she and I, except for the seemingly programmed, 'Okay, I think we should go now' every time we would depart.

She would quiz me on birds and trees, and I never did very well. I had worked in forestry for a little while and my grandma was a birder, and because Diane knew these facts, she always acted playfully disappointed at my poor performance. Nonetheless, I was a quick study and began to retain some of the knowledge she imparted on me. I also started to figure out that she has an awfully dry sense of humour, something near and dear to me. All those hard times she had been giving me came from a loving place. Once I figured this out I would tease her, too; I think it helped her feel young and healthy, even if only for a moment. But once we arrived at the marina, and I pulled up next to the picnic area, she waited patiently for me to help her out and over to the picnic table. She no longer could sit in the low-slung camping chair, and a fluffy pillow had made its way into use instead. Worry still laid on her dog bed from the car, and Diane never truly settled in until that was in place next to her and Worry had laid down on it. The dog could no longer put her head on Diane's lap, so now it usually rested near Diane's leg, almost always touching it.

On a Tuesday in late March, almost five months after our first outing, Diane's conversations with Worry returned to the tone of her first ominous comment that caught me off guard a while back. Diane and I had discussed her plans for the upcoming summer; she was going to spend a couple of months with her family on the beach. This knowledge must have been the impetus behind this statement: 'Worry, I may not see you again. I'll be gone for a while and I don't know if I'll come back.' Worry was laying on her side, eyes closed, with a gentle breeze blowing off the water lightly ruffling her coat. At Diane's mention of her name, though, she looked up, blinked, and closed her eyes again. It obviously probably did not 'mean' anything, but I wanted it to mean, 'No, we will. It's okay.' In reality, though, I just was not sure. After all these months together, I had gotten comfortable in our routine, and I knew I would miss it once it was over. I was certain Worry would too.

Summer breeze

Diane missed her first Tuesday in months at the end of April (well, the first of her doing). I had to cancel some for work obligations from time to time, which she always ribbed me about, but this particular Tuesday she was not up to it. In fact, while we rarely talked by phone unless one of us was running late, I got a call from one of her caretakers saying that she would not be able to leave the house that day. It was instantly saddening, but the caretaker assured me that she would be back the following week. She even told me that, 'Worry can stop by if she's free though.' I am not certain if this is true, but I could have sworn I heard Diane chuckle in the background when this was said. We stopped by that day, briefly. Worry got a lot of love, and gave some back; it was pretty clear they both enjoyed the visit.

The following Tuesday, the first in May, the caretaker called again to say that Diane was under the weather still. I asked, 'Should Worry stop by?' The caretaker replied, 'No, I don't think it will be a good idea today. If something changes, I'll let you know.' That typically reserved hour for driving and time spent at the marina was especially restless that day. It would be safe to say I was 'out-of-sorts.' Worry seemed restless too.

The next day Diane called to say that she was moving up her plans to go stay with her son's family on the beach. She did not need to say more, I could hear it in her inflection; she was not doing well. I told her to write me some letters and we could be pen pals for the summer until she came back. There was a pregnant pause that was thickened by reality settling in. 'I'll see what I can do about that,' said Diane. I responded, 'Okay. Well, Worry would like a picture of the ocean. Take one if you get the chance.' I am not sure why I said that, but I think I was trying to maintain a positive tone, and only something corny would save my voice from cracking. She responded, 'Who's got time for that? Not us dyin' folk!' I laughed. She asked me to stop by with Worry one more time, and I did, the next day. Upon arrival, the pup displayed her usual excitement at seeing Diane: an endearing whine, a helicopter-ing tail that batted everything in its way, and of course, bright eyes that signalled the sight of her 'long-lost' friend. Their exchange was sweet and playful, as much as Diane could muster, and her parting goodbye to Worry was strong and confident. It was almost like she was protecting Worry from the obvious: they likely would not see each other again. Worry, ever ready for the next adventure, trotted through the door on our way out, but she stopped to look over her shoulder; I think she could feel the warm rays of Diane's smile on her fur. I was unable to look back, but because I had seen that exchange before, I could feel it in the air as my arm hairs bristled.

A few days later, a Tuesday, Diane's son picked her up and drove her back to his place at the beach. Several more days past when I realised everything had been moved out of her apartment. She had not mentioned it, but upon reflection it was obvious that her time in this city was through. She would live out her days on the beach with her family, soaking in the warm rays and summer breeze off the ocean, the way she had done with Worry at the marina.

Discussion

Diane's upbringing and a majority of her healthy life were spent in nature and with dogs as faithful companions, thus suggesting a clear importance of both to her life course development (Hendricks, 2012). Forests and other natural environments have been documented as 'therapeutic landscapes' by many (*cf.* Morita et al., 2007), and this can also be especially true for those diagnosed with cancer

(Cimprich & Ronis, 2003). Equally so, as evidenced by the story of Diane, dogs can provide extraordinary levels of comfort for people coming to terms with serious illnesses (Moody et al., 2002). Diane's 'conversations' and time spent with Worry allowed her to externalise her fears and thoughts related to her mortality, thus creating a feeling of catharsis in the process (Marcus, 2012).

For Diane, the synthesis of two meaningful leisure activities, time spent in nature and the company of Worry, granted her coping mechanisms (Hutchinson et al., 2003) that were otherwise not readily available. Since she had not been active in the outdoors in quite some time, and lost her last dog roughly three years earlier, two very important aspects of her identity had been absent from her life. That these significant aspects would resurface late in her life is both humbling and saddening. I have often wondered how both could have been beneficial to her earlier on in the diagnosis.

Diane seemed to find peace out at the marina with Worry; she became comfortable with where she was in life by sharing her introspections with her (Martin & Kleiber, 2005). By having a say in how she found healing through leisure highlights the importance of the holistic care practice for those with serious and terminal illnesses (Hutchinson et al., 2003). And while Diane may not have been looking to have conversations about her illness with me directly, the mere comforting presence of Worry stimulated her ability to release the difficult emotions that were stuck inside (White et al., 2015). Complementary therapies and support are essential to those receiving medicalised care (Docherty, 2004; Marcus, 2012), but for those beyond the treatment stage, the important and necessary support is to be found in the relationships and activities that are of the most significance (Fegg et al., 2010). For Diane, and many others, dogs provide a kind of support that is not found elsewhere (Carr, 2014; Fox & Gee, 2017).

Conclusion

On the fourth consecutive Tuesday that the three of us did not go to the marina together, Worry and I went there without Diane. I tried to sit at the picnic table but it was difficult; it was too different. I never liked doing it before meeting Diane, and I guess with her now several hundred miles away I came around to not liking sitting at the marina again. Instead we went for a hike. Worry seemed happy about that. I was torn. I do not think Worry had forgotten about Diane already, but in some way I interpreted her excitement in bounding through the forest as an extension of her friendship with Diane; I guess it is what Diane would have wanted for her dear friend.

Diane told me before she left that while her health had gotten worse, she did not think she would die anytime soon. She did not indicate whether that was a good thing or not. The most meaningful thing she did tell me was how much she loved my dog and appreciated my willingness to share Worry with her. Her parting words were that I should 'keep taking care of that dog like she is the most important thing in the world.' I agreed to do just that.

Worry clearly provided Diane a lot of comfort, as evidenced by their six months together, and coupled with the natural beauty of the marina, I think we both helped Diane find a little peace with her station in life. While I have seen Worry bond with other humans, there was something special about her interactions with Diane. Worry has an observant and curious personality, but she is often quite reserved around most people. However, from the very beginning it was clear that Worry took to Diane—and vice versa—quite quickly. It was almost as if each of them were kindred spirits who happened to be in need of some extra companionship in their lives. Charles (2016) posited that animals can be surrogates for significant human-human relationships, something that Diane was clearly in need of. Just the same, Diane treated Worry like a close confidant, or an old friend (Evans-Wilday, Hall, Hogue, & Mills, 2018). She confided in Worry, acknowledging her as an important 'person' at a pivotal moment in her life (Taylor, 2007).

As Worry is also getting older, I think maybe there was some psychic connection between the 'two aging ladies' as Diane once referred to herself and her friend. Haraway (2003) stated that the relationships between a dog and their humans are co-constructed, and that these friendships are dependent on the forming of a language that is unique to the establishment and evolution of that bond. As was evidenced, the connection was not a 'given,' but 'created,' between Diane and Worry. Charles (2014) remarked that dogs can provide a sense of 'ontological security' that leads to a reciprocal feeling of kinship between canine and human that is rooted in 'connectedness and belonging' (p. 726). The personal cues that Diane showed and shared with Worry invited the unique traits of her dog-ness to attend to the very specific needs Diane sought: a non-judgmental and open friend to simply spend time with (Evans-Wilday et al., 2018; Sanford, Burt, & Meyers-Manor, 2018)

And now that Diane is not a regular part of our lives any longer, I find myself even more in tune with this dog, something I never thought possible. We have always had wonderful chemistry —I love her like a kid—but now I look at her as the next aging friend I have to make sure has a rewarding last few years of life. She has not started to slow down, yet, but I have become ever more aware that I need to really embrace my time with her as much as possible. In some ways, this obvious point was shown to me indirectly by Diane: nature and close companionship, even between species, are two imperative components of happy lives for many people. I always knew this, but now that I have been reminded of it by someone who came into our lives almost as quickly as she exited, the beauty of these connections between humans and animals has never been more apparent.

Note

1. A pseudonym.

Disclosure statement

No potential conflict of interest was reported by the author.

References

Breault, R. A. (2016). Emerging issues in duoethnography. *International Journal of Qualitative Studies in Education*, *29*(6), 777–794.

Browder, L. M. (2009). Paws and relax. *Home Healthcare Nurse*, *27*(7), 443–448.

Carr, N. (2014). *Dogs in the leisure experience*. Oxfordshire, UK: CABI.

Carter, B., & Charles, N. (2018). The animal challenge to sociology. *European Journal of Social Theory*, *21*(1), 79–97.

Charles, N. (2014). 'Animals just love you as you are': Experiencing kinship across the species barrier. *Sociology*, *48*(4), 715–730.

Charles, N. (2016). Post-human families? Dog-human relations in the domestic sphere. *Sociological Research Online*, *21*(3), 1–12.

Chochinov, H. M., Hack, T., Hassard, T., Kristjanson, L. J., McClement, S., & Harlos, M. (2005). Understanding the will to live in patients nearing death. *Psychosomatics*, *46*(1), 7–10.

Chochinov, H. M., Hassard, T., McClement, S., Hack, T., Kristjanson, L. J., Harlos, M., ... Murray, A. (2009). The landscape of the terminally ill. *Journal of Pain and Symptom Management*, *38*(5), 641–649.

Chubak, J., Hawkes, R., Dudzik, C., Foose-Foster, J. M., Eaton, L., Johnson, R. H., & MacPherson, C. F. (2017). Pilot study of therapy dog visits for inpatient youth with cancer. *Journal of Pediatric Oncology Nursing*, *34*(5), 331–341.

Cimprich, B., & Ronis, D. L. (2003). An environmental intervention to restore attention in women with newly diagnosed breast cancer. *Cancer Nursing*, *26*(4), 284–292.

Creagan, E. T., Bauer, B. A., Thomley, B. S., & Borg, J. M. (2015). Animal-assisted therapy at Mayo clinic: The time is now. *Complementary Therapies in Clinical Practice*, *21*, 101–104.

Crossman, M. K., Kazdin, A. E., & Knudson, K. (2015). Brief unstructured interactions with a dog reduces distress. *Anthrozoös*, *28*(4), 649–659.

Denzin, N. (2006). Analytic autoethnography, or déjà vu all over again. *Journal of Contemporary Ethnography*, *35*, 419–428.

Docherty, A. (2004). Experience, functions and benefits of a cancer support group. *Patient Education and Counseling*, *55*, 87–93.

Duvall Antonacopoulos, N. M., & Pychyl, T. A. (2014). An examination of the possible benefits for well-being arising from the social interactions that occur while dog walking. *Society & Animals*, *22*, 459–480.

Ellis, C. (1998). Exploring loss through autoethnographic inquiry: Autoethnographic stories, co- constructed narratives, and interactive interviews. In J. Harvey (Ed.), *Perspectives on loss: A sourcebook* (pp. 49–62). Philadelphia: Brunner/Mazel.

Ellis, C. (2004). *The ethnographic I: A methodological novel about teaching and doing autoethnography*. Walnut Creek, CA: AltaMira.

Engleman, S. R. (2013). Palliative care and use of animal-assisted therapy. *Omega*, *67*(1–2), 63–67.

Evans-Wilday, A. S., Hall, S. S., Hogue, T. E., & Mills, D. S. (2018). Self-disclosure with dogs: Dog owners' and non-dog owners' willingness to disclose emotional topics. *Anthrzoös*, *31*(3), 353–366.

Fegg, M. J., Brandstätter, M., Kramer, M., Kögler, M., Haarmann-Doetkotte, S., & Borasio, G. D. (2010). Meaning in life and palliative care. *Journal of Pain and Symptom Management*, *40*(4), 502–509.

Fox, R., & Gee, N. R. (2017). Great expectations: Changing social, spatial and emotional understandings of the companion animal-human relationship. *Social & Cultural Geography*, 1–21. doi:10.1080/14649365.2017.1347954

Gubrium, J. F., & Holstein, J. A. (2008). Narrative ethnography. In S. N. Hesse-Biber & P. Leavy (Eds.), *Handbook of emergent methods* (pp. 241–264). New York: Guilford Press.

Hamilton, L., & Taylor, N. (2017). *Ethnography after humanism: Power, politics, and method in multi-species research*. Basingstoke, UK: Palgrave MacMillan.

Haraway, D. (2003). *The companion species manifesto: Dogs, people, and significant otherness*. Chicago: Prickly Paradigm Press.

Harmon, J. (2018). Celebrate the trail to recovery: power of the positive post-diagnosis of cancer. *International Journal of the Sociology of Leisure*. doi:10.1007/s41978-018-0014-x

Hendricks, J. (2012). Considering life course concepts. *Psychological Sciences and Social Sciences*, *67*(2), 226–231.

Horowitz, S. (2010). Animal-assisted therapy for inpatients: Tapping the unique healing power of the human-animal bond. *Alternative and Complementary Therapies*, *16*(6), 339–343.

Hutchinson, S. L., Loy, D. P., Kleiber, D. A., & Dattilo, J. (2003). Leisure as a coping resource: Variations in coping with traumatic injury and illness. *Leisure Sciences*, *25*, 143–161.

Irvine, L. (2004). A model of animal selfhood: Expanding interactionist possibilities. *Symbolic Interaction*, *27*, 3–21.

Janke, M. C., & Jones, J. J. (2016). Using leisure to find a way forward after loss: A lifespan perspective. In D. A. Kleiber & F. A. McGuire (Eds.), *Leisure and human development* (pp. 293–317). Urbana, IL: Sagamore.

Johnson, R. A., Meadows, R. L., Haubner, J. S., & Sevedge, K. (2008). Animal-assisted activity among patients with cancer: Effects on mood, fatigue, self-perceived health, and sense of coherence. *Oncology Nursing Forum*, *35*(2), 225–232.

Kleiber, D. A., Hutchinson, S. L., & Williams, R. (2002). Leisure as a resource in coping with negative life events" Self-protection, self-restoration, and personal transformation. *Leisure Sciences*, *24*(2), 219–235.

Krause, N. (2009). Meaning in life and mortality. *Journal of Gerontology: Social Sciences*, *64B*(4), 517–527.

Kulick, D. (2017). Human-animal communication. *Annual Review of Anthropology*, *46*, 357–378.

Marcus, D. A. (2012). Complementary medicine in cancer care: Adding a therapy dog to the team. *Current Pain and Headache Reports*, *16*, 289–291.

Marcus, D. A., Bernstein, C. D., Constantin, J. M., Kunkel, F. A., Breuer, P., & Hanlon, R. B. (2012). Animal-assisted therapy at an outpatient pain management clinic. *Pain Medicine*, *13*, 45–57.

Martin, L. L., & Kleiber, D. A. (2005). Letting go of the negative: Psychological growth from a close brush with death. *Traumatology*, *11*(4), 221–232.

Moody, W. J., Maps, R. K., & O'Rourke, S. (2002). Attitudes of paediatric medical ward staff to a dog visitation programme. *Journal of Clinical Nursing*, *11*, 537–544.

Morita, E., Fukuda, S., Nagano, J., Hamajima, N., Yamamoto, H., Iwai, Y., … Shirakawa, T. (2007). Psychological effects of forest environments on healthy adults: Shinrin-yoku (forest-air bathing, walking) as a possible method of stress reduction. *Public Health*, *121*, 54–63.

Nakau, M., Imanishi,Ji., Imanishi, Ju., Watanabe, S., Imanishi, A., Baba, T., Hirai, K., Ito, T., Chiba, W., Morimoto, Y. (2013). Spiritual care of cancer patients by integrated medicine in urban green space: A pilot study. *Explore: The Journal of Science and Healing*, *9*(2), 87–90.

Norris, J., & Sawyer, R. (2012). Toward a dialogic method. In J. Norris, R. Sawyer, & D. Lund (Eds.), *Duoethnography: Dialogic methods for social, health, and educational research* (pp. 1-32). Walnut Creek, CA: Left Coast Press.

Sanford, E. M., Burt, E. R., & Meyers-Manor, J. E. (2018). Timmy's in the well: Empathy and prosocial helping in dogs. *Learning & Behavior*. doi:10.3758/s13420-018-0332-3

Smith, C. M., Treharne, G. J., & Tumility, S. (2017). All those ingredients of the walk: The therapeutic spaces of dog-walking for people with long-term health conditions. *Anthrozoös, 30*(2), 327–340.

Song, C., Ikei, H., & Miyazaki, Y. (2016). Physiological effects of nature therapy: A review of the research in Japan. *International Journal of Environmental Research and Public Health, 13*, 781–798.

Taylor, N. (2007). 'Never an It': Intersubjectivity and the creation of animal personhood in animal shelters. *Qualitative Sociology Review, 3*(1), 59–73.

Tomás-Sábado, J., Villavicencio-Chávez, C., Monforte-Royo, C., Guerrero-Torrelles, M., Fegg, M. J., & Balaguer, A. (2015). What gives meaning in life to patients with advanced cancer? A comparison between Spanish, German, and Swiss patients. *Journal of Pain and Symptom Management, 50*(6), 861–866.

Tracy, S. (2013). *Qualitative research methods: Collecting evidence, crafting analysis, communicating impact.* Hoboken, NJ: Wiley-Blackwell.

Urbanski, B. L., & Lazenby, M. (2012). Distress among hospitalized pediatric cancer patients modified by pet-therapy intervention to improve quality of life. *Journal of Pediatric Oncology Nursing, 29*(5), 272–282.

Warne, K. E., & Hoppes, S. (2009). Lessons in living and dying from my first patient: An autoethnography. *Canadian Journal of Occupational Therapy, 76*(4), 309–316.

White, J. H., Quinn, M., Garland, S., Dirkse, D., Wiebe, P., Hermann, M., & Carlson, L. E. (2015). Animal-assisted therapy and counseling support for women with breast cancer: An exploration of patient's perceptions. *Integrative Cancer Therapies, 14*(5), 460–467.

Yong, E. (2015, October). A genetic study writes a new origin story for dogs. *The Atlantic.* Retrieved from https://www.theatlantic.com/science/archive/2015/10/genetic-study-writes-yet-another-origin-story-for-dogs/411196/

Sport horse leisure and the phenomenology of interspecies embodiment

Andrea Ford ⓘ

ABSTRACT
This article presents an auto-ethnography of the experience of sport horse riding. Drawing on phenomenological and anthropological theories of embodiment, I argue that the aspirational goal of sport riding is co-embodiment between horse and human, in which kinesthetic perception, intention, and volition merge. Co-embodiment requires time and practice to develop a shared multi-species culture in which bodies can be attuned to one another, and profound attention to both the immediate moment and the other being. I suggest that the interspecies component of sport riding, and the sport component of the interspecies engagement, is a significant part of what makes it appealing as a leisure activity. It invites (and requires) an experience of corporeal immediacy and intimacy that is both deeply satisfying and absent from many work or social environments in contemporary, wealthy, Western societies. Methodologically, this article draws from two decades of personal experience in sport horse riding, and engages auto-ethnography as a way to use the researcher's body as phenomenological tool while keeping broader cultural contexts in mind.

Introduction

Leisure is something other than the demands of work, and in some cases other than the demands of social life. Sport horse riding has been a favorite leisure activity of mine for two decades now, and has many times transgressed the border with both work and socializing, as I was paid to teach lessons, coach competitors, do barn chores, and train and exercise other people's horses, and maintained friendships and social networks in the equestrian world. Yet its draw has always been something other than payment or interpersonal connections. By exploring what I find to be the most compelling aspect of riding, which I call interspecies co-embodiment, I discuss the unique role equestrian sports play in leisure culture. In this auto-ethnographic paper, I draw from philosophy and anthropology to think through the phenomenology and cultural relevance of intentional co-embodiment with a non-human, arguing that the immediacy and intimacy such connections facilitate play a role in the pleasure of the sport as leisure.

Achieving interspecies co-embodiment is the result of long practice and keenly developed intuition. Although often fleeting and aspirational, it is an experience of profound immediacy that I experience as an antidote to the distraction and anxiety that permeates the information-based academic world where I otherwise work. One must be 'present' to connect on this level with an animal. Sharing embodied awareness across species also provides a bodily intimacy that is not generally found in social or individual leisure pursuits or work endeavors. Both this immediacy and intimacy are enabled by, and require, embodied attention. I suggest that engaging with another species facilitates and enhances our capacities for embodied attention, and that this is a key component of sport riding as leisure. Co-embodiment is a theme in my primary ethnographic research on childbearing, including pregnancy, birth, and infant care, and my other leisure pastime of partnered dance, though in neither of these contexts is it interspecies. In this paper I highlight what makes the interspecies component of co-embodiment unique and important in constituting sport riding *as* leisure. Ultimately, I suggest that the role of horses in leisure landscapes serves to legitimate and encourage attention to corporeal experience in ways that are absent from many 'work' environments, as well as many human social ones.

Method

Sport riding typically includes the Olympic disciplines of dressage, eventing, and show jumping, which are the styles of riding in which I have experience. The category could easily be extended to include other traditions, from endurance riding to Western cutting or reining. For the purposes of this article, the key aspect of 'sport riding' is that horse and rider are united to accomplish an activity that is physically and mentally challenging for them both. In this paper I draw from my equestrian experience in Northern California and the Chicago area over two decades (1998–2018), from beginner to quasi-professional. During this period I was involved with ten different barns, took and taught lessons, competed locally and regionally, and coached students at shows. I exercised and trained countless horses, including two that I owned and two that I leased, started several young horses who had never been ridden, and spent a few summers as a 'working student' (similar to an apprentice or intern). Generalizations drawn from my experience necessarily speak to wealthy Western societies and the United States in particular. My conception of sport riding is closer to the 'natural horsemanship' that Latimer and Birke (2009) distinguish from 'traditional rural communities' in the UK, in which the former seek a dyadic and mutual relationship with minimal coercion or force, and the latter approach equine sport and competition as a way of life. However, in the US the term 'natural horsemanship' connotes a specific cultural community distinct from sport riding; my riding practice might better be described as 'ethical equitation' (McLean & McGreevy, 2010) based on traditional European training principles.

Auto-ethnography is a particularly suitable method for making these claims. To understand how riding *is* (phenomenologically) and therefore how it can be understood as co-embodiment, it must be experienced. In positing a theoretical framework and methodology for describing and analyzing embodied life, Deidre Sklar (1994) claims that the researcher's own body is a point of access to corporeal knowledge in cultural practice. She insists that such research must be practiced; it is not enough to observe, because the embodied experience of something yields more than 'meets the eye'. Although vision is the privileged form of experience in the West, much somatic experience exceeds the visual, or even the perceptual. Sklar writes, 'It is a matter of re-cognizing kinesthetically what is perceived visually, aurally, or tactilely' (p. 15). My extensive kinesthetic experience of sport riding enables me to understand the claims I am presenting on a deep corporeal level, which is in fact what enables me to make them. If I were working from interviews, I would likely be able to piece together similar understandings of the experience of riding, but being the theorist *and* the body in question offers a particularly poignant perspective.

Deborah Butler (2017) has advocated a similar 'autophenomenography' to apply phenomenology within the sociology of horse racing as both a methodological approach and a theoretical framework, focused on 'the researcher's own lived experience of a phenomenon' (p. 125; see Allen-Collinson 2011 for a more general use of the term in sports analysis). I use 'auto-ethnography' instead because, in addition to being a more approachable term, it is less constrained to phenomenology and more engaged with broader anthropological discussions of culture. Thomas Csordas' (1993) conceives of embodiment and somatic attention as an essential methodological complement to semiotic analyses of culture-as-meaning. He subsumes phenomenological approaches within 'embodiment' as a methodological field, arguing that 'perceptual experience and the mode of presence and engagement in the world… is the starting point for analyzing human participation in a cultural world' (p. 135). For him, embodiment is 'a dialectic between perceptual consciousness and collective practice (p. 137). Because my goal is to draw from phenomenology in order to theorize culture, including from a multi-species perspective, thinking in terms of ethnography is more appropriately capacious. The italicized descriptions below are composites of my memories and serve as auto-ethnographic 'data'. This approach has obvious limitations in terms of generalizability, as innumerable personal histories and qualities shape my experience. More generally in qualitative methodology, auto-ethnography has been critiqued for this narrowness and potential for self-absorption, yet advocated as a more honest, artistic, enjoyable relationship to the personal experience that qualifies all research (Ellis, 1999). Relevant to my purposes here, Sklar asserts that 'The insights gained from a corporeal approach derive from an emphasis on depth rather than breadth, a stripping down rather than a piling up of associations' (1994, p. 18). This has guided my approach; I aim to describe and theorize the essence of how riding is for me, one particular culturally-embedded person, in moments of seamless connection between species.

Embodiment and attention

I open the car door and crunch onto the weedy gravel. It sounds so different from pavement. The silence here is not quiet; it is speechless, but brimming with other sound. The distant whooshing of traffic seems contained in a different world. Cows mull over their cud, shifting their glossy black hides highlighted with chestnut fluff. The pattern of their dark bodies sprawls over the variegated grass that is patchy with yellow, purple, peridot, rust… a rich, fresh green that is anything but, rippling exactly as if it were some long-haired creature stroked by the wind. It is idyllic behind the barbed wire. I climb the towering flight of concrete stairs to the barn, boots producing a solid, sturdy version of high-heels clicking. Hugging my grungy sweatshirt close against the sharp wind, I watch my calves encased in leather chaps, shiny spurs buckled around my ankles, looking armored and confident. The barn smells of dust, wood shavings, and alfalfa, of fur and manure and faintly of leather. Musty, yet alive. The mobile scent of outdoor air combines with the aromatic heat of stationary animal bulk. Their bodies are hidden in the obscurity of their stalls but their warmth is tangible in the air nonetheless. They mill in circles, stomp, chomp, occasionally whicker while I click-clomp up the paved aisle, tense against the chill.

I grab a halter; the tacky leather is supple and molds into my palm, its buckles dull, the heavy cotton rope draped over my shoulder. Horses glance as I pass, sometimes gloomily, their bodies turned away and sullen, sometimes perkily, eagerly swiveling forward their ears. Ivanna has already perked up and turned towards me by the time I slide open the clanging, grating door of her stall. Seeing her dark familiar face with its bright star, massive head reaching forward with fond attention, causes a soft pang of gratitude in my chest. I rub her cheekbones and feed her a carrot. The deeply bedded shavings sink under my feet, my steps poufing up whorls of sawdust that tickle my nose and make me sneeze. Folding back her blanket and sliding it off releases welcome heat, and I wonder if she is cold now. I do not ask her permission as I strap her head into the leather halter, cold fingers fumbling with the buckles, burrowing backwards for warmth in my sleeves once the task is accomplished. Yet there is a subtle kind of communication, an assertive assumption on my part, unhesitant yet attentive to her consent. This is wedded to domestic habit; Ivanna was feral when we first met, and her habituated attitude was cultivated over many months, turned into years.

Cross tied in the aisle for grooming, Ivanna fidgets. She tosses her head to make the chains clank. I brush her, flicking up tiny poufs of dust and scattered shavings that were trapped under her blanket, and her flesh is smooth and firm, the muscle taut in places and fluid in others. My movements are hurried to warm me up, and my muscles are auto-pilot proficient at the grooming routine. I lean against her flank to shift her weight off the foot I'm about to pick up. Its mass rests heavily in my palm; the texture is much like a smooth rock, but that comparison never suggests itself because the hoof is warm, lively with the propensity to motion and growth. I set it down and can sense Ivanna shift her weight to prepare the next foot for picking up and picking out. I am attuned to her body with a subtle spatial awareness, something other than the slight visual modification or near-imperceptible grinding against the rubber mat.

Grooming finished, I run my hand over her muzzle as I walk to the saddle stand across the aisle. She blows an exaggerated breath that's not quite a snort, moist and visible in the early chill, the faint odor of chewed hay mingling with a plethora of kindred scents. I am attentive to the slight flicks of her ears and what they indicate, ready to be instantaneously responsive to any tension in her carriage. She was quite explosive when feral. I am careful to make sure she is aware of what I'm doing and tacitly consenting to it, so she isn't startled and dangerously indignant at the surprise. But despite this attention, there is not the self-consciousness I exercise with a human, the awareness of how whatever relationship we have or don't have with each other conditions our interaction. Here, my actions are functional, somehow straightforward, and I have the confidence of routine and tacit authority – though complicated by her unpredictability. I do what I am doing regardless of what Ivanna 'thinks of me', but am highly aware of what she thinks of the situation. I buckle on the saddle, hushing her in a low monotone as the girth's pressure increases, which has been a point of reactivity in the past. Do you say that a horse has trauma? Is that where her sensitivity to being cinched round the belly comes from, some event in her history? Or is it 'just' unpleasant? Can you disentangle bodily sensation from memory? I reach up around her head and draw up the bridle, sliding the bit between her big teeth to the bars of her gums, which she doesn't mind in the slightest.

***

Sport horse riding, which is fundamentally a multi-species activity, occupies a border-zone between individual and social activities that serve as leisure pursuits. As complex animals, horses have a subjectivity with which riders engage, but this subjectivity is less (or at least differently) demanding than that of other humans. I do not mean to imply that horses are some kind of midway point between object and subject, though this type of language is common to Western cultural conceptions of animals (Haraway, 2008, p. 206). Rather, I am making the more specific observation that the attention riders pay horses has to account for their consciousness and alterity but not the layers of social conventions, etiquette, judgments, and abstracted meanings that overlay human interactions, and their consequent stakes. The process of learning how to pay attention to a horse can be thought of as an ethical endeavor to the extent that it is premised on an individual horse's intrinsic worth as a sentient being, an approach that fosters a mutually rewarding partnership (Dashper, 2017).

Horses require attention, but of a deeply corporeal nature. On one level, this involves 'body language'. Aspects of body language that are particularly interesting, but outside the scope of this paper, include the ways horse handlers 'translate' equine herd 'language' into their own bodily comportment (Birke, 2008), and the way 'confident' body language is embedded in gendered appreciations of horsemanship (Brandt, 2005). Rather, this paper focuses on a different level of corporeal attention that exceeds linguistic metaphors. When actively engaged in sport, as we will see below, such attention can be refined into a co-embodiment that bypasses the need for semiotic communication. The rider and horse share a sense of their combined embodied action in the world. Such refinement is the goal of high-level sport riding.

Co-embodiment is an aspirational ideal that is only achieved with regularity by highly skilled practitioners. However, beginners' fleeting experiences of co-embodiment can be highly motivating, as was the case in my own experience, reinforced through subsequent years of teaching lessons punctuated with joyful exclamations of 'oh, that's what it's supposed to feel like!' When learning or teaching the basics of riding, linguistic metaphors are common. Pulling on the reins in such a way 'says' halt or turn, kicking at such a moment 'tells' the horse to not stop, etc. As the rider's skill level rises, instructions from the trainer often become about the rider's body without necessarily specifying the 'message' intended by the bodily shift: 'look up' or 'let your leg hang' for example. Most riders will come to understand some kind of theory behind human motions and equine responses, but this dynamic becomes more and more implicit. Looking well in advance towards one's next jump prepares the horse to arrive there in balance; drawing in one's abdomen and lengthening one's spine in a 'half-halt' results in the horse lifting and rounding its back. The theory behind advanced techniques becomes less and less semiotic (x *means* y) and more and more sympathetic, attuned, automatic (x *is* y). I argue that this is because horse and rider come to share a sense of embodiment.

The kind of attention that yields this co-embodiment is both learned and interactive. I build on Csordas' concept of 'somatic modes of attention', which he defines as 'culturally elaborated ways of attending to and with one's body in surroundings that include the embodied presence of others' (1993, p. 138). Bodily modes of attention are always culturally constituted – i.e. learned, neither biologically determined nor arbitrary. The long process of refining horse-human 'body language' into co-embodiment is a cultural process in which both horse and human participate (on human-animal cultural creation, communication, and knowledge production, see Haraway, 2008; Smart, 2011). Perception is always embedded in a cultural world, and horse and human learn to perceive each other. As to the second point, that attention is interactive, Csordas builds on Merleau-Ponty's (1962) observation that attention to something actually brings it into being. Attention is the 'existentially ambiguous point' where the object of the attention both exists as itself and as an interaction with what is attending to it (Csordas, 1993, p. 138). Likewise, the object of attention shapes both the quality of the attention and the person doing the attending. Attention is always both 'to' and 'with'. It always occurs in the present moment, and indeed brings the present moment into being.

Embodiment and union

The stirrup leathers snap as I run down the irons. I cluck and gently tug on Ivanna's reins; she steps forward and halts next to the mounting block at my 'hup'. Although an observer wouldn't notice much but a smooth set of motions, I am highly attuned to Ivanna's potential action. Mounting was a primary difficulty when training her; she bucked me off before I ever sat on her back! It is difficult to even describe what 'signals' I am 'looking for' – she is now so accustomed to mounting that there are far fewer overt indications: stepping sideways, back tensing, lifted head and rolling eye. This required long patience and calm repetition from me. Now, it is more that I have a habituated readiness to notice, a continued calm attention that likely reassures her. All goes smoothly and I settle into the saddle. Riding comes in intense chunks for me these days; my sore inner thighs feel the subtle knob of the stirrup bar, and I know my knees and ankles will soon complain about the extra flexion and concussion. Yet the saddle still feels like home.

She walks. Although I am adjusting my reins, stretching, aligning, positioning my feet, I am still ready. If she spooks at one of the swooping hawks, I will not fall. The physical qualities of a rider are subtle – I may perhaps look like someone sitting upon a horse's back, as a beginner is, but my balance is 'on', my reaction time is quick, and the reaction is the appropriate one. I am 'riding' even when I am just sitting there. I am ready, which requires both the skill of long practice and keen attention in the moment. To move from walk to trot, my calves squeeze just until she transitions, and then release, as my hip angle closes and my balance shifts slightly forward. I do not usually think about any of this though. When we are warmed up and synchronized, I simply think 'trot' and it all happens beneath conscious awareness. My legs effortlessly coordinate with the motion of her back to lift me into 'posting', my rise correctly synchronized with her outside leg moving forward, which creates more thrust than the inside leg as it travels further on the circumference of our curved path. Ages ago, my twelve-year-old self shed frustrated tears over the impossibility of finding this rhythm.

Sometimes while riding my thoughts are elsewhere. I can go on 'auto-pilot' and rely on physical memory while my 'mind' wanders. But this is simply habit, not co-embodiment. Co-embodiment requires attention, and it is not just the rider who has to pay attention, either. Ivanna is like a diesel truck, I tell people. She takes a very long warm up before her engine is primed and self-sufficient. Now that she is advanced so far into domesticity and no longer on the feral edge, getting her to pay attention can be trying. 'Motivate, don't nag!' trainers tell me. Easier said than done. But when she is fit and fluid her legs do really seem like pistons, generating energy that loops up from her hind hooves through her back, through me, down through her neck and bit, and back up again. I feel like this energy is somehow contained between my legs and hands, in me as well as in her, a force comprised of us both, by which we are both carried along. I can hear her piston-rhythm accelerating, the trot collapsing into a run, energy loop broken as her front feet pound, dissipating it into the ground. I lift my bellybutton up into my spine, roll back my shoulders and tuck my pelvis, lifting her 1500 pound weight back onto her haunches, generating more energy through pulsing squeezes of my calves, bringing the energy back into a loop. I lift my inside rein – maybe not even an inch, maybe just the thought of it – to keep her vertical balance around the curve, hold her there upright between my lifted left hand and supporting right leg, and just as her right hind hoof lifts I press my right heel towards her belly, and her hoof lands in the first footfall of the canter. From two beats to three: waltz tempo.

Ivanna and I canter, and it feels like we are an enormous featherweight rocking chair. That is, until the pieces fall apart again, to be rebalanced, soon to be distracted, soon to be re-attuned, constantly in the process of refinement. A rider must have muscular power in the legs, the arms, the torso, but her real physical ability is being strong in subtlety, in holding herself perfectly in balance throughout all the intense motion of galloping and leaping and prancing sideways, appearing still. My gaze softens to the middle-distance as I tune-in to Ivanna's footfalls, her alignment, the quality of her energy. Her forward motion flags and the arc of her spine becomes wiggly as her energy threatens to flop out to the sides; a nudge with my seat and tap with both ankles drives her forward and straight. She pounds the ground and it pushes back; I feel the pounding course through my ankles in rhythm with her regular snorty breath, and my own breath is controlled and rhythmic as well. I lift my seat out of the saddle and encourage her to open her stride into a gallop. I am aware of my lower back muscles holding me upright, and I remind myself not to compress my left side in my inclination to asymmetry. My head turns to calibrate the corner, automatically adjusting the relative balance of our bodies' alignment to make her speed coincide with the calibration. It is at once an experience of rapidity as the fence posts flash by and her staccato hoof beats become indistinct, and of tranquility as my gaze takes in the panoramic countryside, the hawk floating lazily at our pace. I roll back my shoulders, open my hip angle, and sink my weight into my heels, closing my fingers around the reins. We come back to a walk, though I am careful to keep my body active so Ivanna doesn't simply 'collapse' into the slower gait. Attunement is a never-ending process. The scent of sweat rises, mingling with dust, and I relax my body and release the reins, allowing Ivanna to stretch her neck down and snuffle the sand.

***

At what point does the horse and rider bodily conversation and negotiation become co-embodiment? Is it a spectrum of subtlety? An open-ended progress? Perhaps. What is clear to me is that we *can* become so engaged in what we are attempting to do together that communication is refined to the point of seamless connection, if only for a brief moment. Co-embodiment requires paying focused attention: immediacy. Such attention, if mutual, enables a unified engagement with the world: intimacy. In the intimacy and immediacy of that connection, we transcend all the detailed, sensitive work taking place between our bodies. Co-embodiment happens when the cues by which the rider signals her intention become so subtle they are subconscious, and her awareness of the horse's intentions becomes so subtle that she subconsciously anticipates his actions, rather than reacting to them. I will not attempt to project onto the specifics of the horse's experience, but he must likewise be subtly engaged.

While horse and rider are often thought of as partners (Birke, 2008; Wipper, 2000), the (albeit fleeting) experience of co-embodiment doesn't seem accurately described as partnership or team-work, which evokes one consciousness communicating across a gulf to another consciousness so they can direct their respective bodies toward a shared goal. Co-embodiment builds on a learned relationship, based on mutual respect, trust, confidence, and close communication as well as compatibility (Wipper, 2000), but then transcends it. In such moments, I know where each of my horse's feet are with the same proprioceptive sense that enables awareness of my own feet, without 'thinking' about it. I may make conscious technical decisions about direction or speed, as I would while walking or running, but I communicate them subconsciously to my horse, as I would communicate them to my own legs. It is a union.

What is a sense of embodiment, such that it can be shared? The concept of a body schema (sometimes also called body image) is one way to theorize this. Elizabeth Grosz describes the body schema as 'an anticipatory plan of (future) action in which a knowledge of the body's current position and capacities for action must be registered' and thus 'formed out of the various modes of contact the subject has with its environment through its actions in the world' (Grosz, 1994, pp. 67–68). Body schema is the link between perception and action; it is, in effect, the perception of one's possibilities for interacting with the world. It allows actions to happen without any transfer between 'mind' and 'body'; the two are co-constituted as the self is engaged in the world. The body schema is learned over time, based on habits of perception and action. This is why sharing a body schema is only possible in expert horse-rider interactions, where both parties have long experience with the other's capacities and likelihoods.

The body schema is responsible for the 'phantom limb' that affects amputee patients who continue to perceive and take action with a part of their body that is missing physically. Conversely, the success of a prosthesis depends on the body schema literally incorporating the new piece of body. Prosthetics can be considered expansively to include not just substitutes for 'defects' (such as artificial limbs) but anything that augments bodily capacities (Grosz, 2005). When one's body schema includes a pair of habitually-worn glasses, one allows extra space for their edges when maneuvering one's head. A novice bicyclist has to think of making particular motions, say turning the handlebars to avoid an obstacle, but when cycling becomes automatic one fluidly moves past obstacles as a unit with the bicycle. If this tool is used to differently enable one's body, if it is incorporated into one's assumed-without-conscious-thought possibilities for interacting in the world, is it not then a part of the person? Jean-Paul Sartre answers: 'my body always extends across the tool which it utilizes: it is at the end of the cane on which I lean and against the earth; it is at the end of the telescope which shows me the stars...' (1984, p. 428). One can subsume parts of the world into one's conception of self, expanding one's capacities, blurring one's edges (Stone, 1994).

But a horse is not a tool. Although Grosz proposes that prosthetic objects may be transformed alongside the body that incorporates them in a 'mutual metamorphosis', provocatively suggesting more agency than typically granted to these inanimate objects (2005, p. 148), a complex mammal like a horse has a consciousness that is relatively easily recognizable by humans. Extending one's body schema to incorporate a horse is much more elusive than for a bicycle. The metaphor of the prosthetic has to become more complicated, as this 'object' fairly explicitly 'talks back' through its own voluntary motions in the world. For Aristotle, motion was the primary condition for consciousness and selfhood, not thought as in the Cartersian legacy; the body's 'animation' by *anima*, or spirit, was the pre-condition of its existence (Sheets-Johnston, 1999). Sharing a body schema with a horse requires not only disciplined practice and skill-learning that create cultural habits and quick reactions, not only a focused bodily attention, but sensitivity and empathetic connection with another spirit. Again, this builds on having established a partnership, but exceeds it. Animal science research has shown what any horse person knows: not only can horses remember humans in ways that impact their subsequent interaction (Fureix, Jego, Sankey, Hausberger 2009), a bond emerging from successive interactions that set 'positive' or 'negative' expectations (Hausberger, Roche, Henry, & Visser, 2008), but emotions like anxiety are communicated between human and equine bodies in real time when executing even simple tasks together (Keeling, Jonare, & Lanneborn, 2009). Grosz notes that the body schema is 'also comprised of various emotional and libidinal attitudes' and that 'the subject's experience of its own body is connected to and mediated by others' relations to their own bodies and to the subject's body' (1994, p. 67–68).

In her description of emotional intersubjectivity with the racehorses she rides, Deborah Butler builds on Merleau-Ponty's (2003) idea of a human-animal 'intertwining', a lateral relationship of overcoming or transgression based on 'the body's openness of being, its ability to touch and be touched, its ability to be both perceiver and perceived, and to communicate nonlinguistically' (Butler, 2017, p. 129). This intertwining of sentient beings, or primordial intercorporeality, is 'a shared belonging of the reversible flesh of the world... a reversibility between myself and an other... an "intentional tissue", a meaningful flow of sensual and haptic communication' (p. 132). She describes how this profound connection can be a catalyst for recovering from emotional imbalance and poor mental health. However, she says 'What we do not achieve, however close and embodied our relationship with an other might be, is to meld into one. Our experiences lived through our bodies are our own, there is an irremediable space between us' (p. 127). While I agree that no durable melding can occur, I argue by contrast that moments of profound connection such as she herself describes can indeed be described as union.

Other theorists have described such an embodied-emotional union. Beth Seaton (2013), referencing Merleau-Ponty (1968), discusses how communications 'amongst the actors of human and non-human worlds involve sensate, kinaesthetic and mindful exchanges of identity. They involve the extensions of body and thought in which emotional and sensory registers act in harmony as material and mindful orientations, allowing the anticipation or intuition of the future movements and emotions of an other which, in some form, may also become like-minded'. She illustrates with an evocative description of riding:

> When my horse and I are riding in balance with one another (which we must be), we share a synchronicity of emotion, perspective and movement. When I am taking him through a jumping course, I must look forward towards our immediate destination as he would look forward (looking at the ground is a sure means of falling to the ground); I must measure the distance and count his strides to the next jump and in counting his movements to myself I am communicating to him how many strides he needs to take (and he does). I need only think about these movements we will next perform, and if I am successful in my thinking, if I am successful in becoming like him, my horse will perform them. If, on the other hand, I am feeling out-of-sorts, if I am stressed or anxious, my horse will react in kind and become tense and spooky. If I miss our count of strides, my horse may, quite possibly, crash into the fence. Similarly, if my horse is overly excited or fearful, then I can have great difficulty in finding that necessary muscular-emotional connection and will have to work very hard at communicating a loose and easy body and mind in order for him to relax.

Vinciane Despret refers to this affinity between horse and rider as 'iso-praxis' (2004, p. 115), a type of body and mind mimicry in which not only do horses respond to their riders, but talented riders behave and move like horses. The riders 'have learned to act in a horse-like fashion, which may explain how horses may be so well attuned to their humans, and how mere thought from one may simultaneously induce the other to move. Human bodies have been transformed by and into a horse's body' (p. 115). Intentions become intertwined, and both human and horse are cause and effect of each other's movements. 'Both embody each other's mind' (*ibid*).

Using the centaur as a mythic figure or archetype accessible to humans on an unconscious level, Ann Game (2001) likewise proposes 'that we are always already part horse, and horses, part human... Through interconnectedness, through our participation in the life of the world, humans are always forever mixed, and thus too have what could be described as a capacity for horseness' (1). She describes how 'When a horse moves freely, balanced, with cadence and lightness, it feels like floating and flying. Ecstatic. These moments of effortless airy floating, flying, so light yet requiring perfect placing of the hooves on the ground... are the moments we ride for' (3). What might be called the 'centaur effect' is the 'ecstatic quality of good riding' (*ibid*), a spiritual exhilaration, 'a feeling that energies have been set free in us that are superior to those that are ordinarily at our command' (*ibid*). Social theorists and their informants describe feelings of oneness during riding as 'mutual becomings' (Oma, 2007), and use terms that evoke telepathy or metaphysics, a transcendent connection with something larger than the self (Brown, 2007). In

England a secret 'horseman's word' was said to convey powers over horses when whispered in their ears: it was 'both in one' (Evans, 2008, p. 246).

In her beautiful inquiry into why such esoteric, mystical descriptions of human-horse interaction are remarkably widespread across time and geography, archaeologist Gala Argent (2012) begins with nonverbal communication as a profoundly important capacity of both horses and humans that is too-often discounted as secondary to human verbal capacities. Both species' ability to communicate nonverbally amongst ourselves (intra-specifically) is what allows us to do so between species (inter-specifically), she claims. Both haptics (touch) and proxemics (spatial positioning) are essential to horse socialization, and they cultivate affection and bonding with particular individuals. Argent suggests horses have a desire to please, and a superior ability to assess intentionality, which allows them to anticipate a rider even before being 'asked.' Quoting Maxine Sheets-Johnstone (2002), she adds that humans, for our part, 'engage in corporeal-kinetic sense-makings' to create an intercorporeal world of common understandings (Argent, 2012, p. 114). Horse herds move as one organism, and draft teams and mare-foal dyads often synchronize footfalls; this capacity for corporeal synchrony extends to horse-human dyads. Argent uses anthropologist Edward Hall's idea of entrainment (1983) to focus on the internal processes that make manifest observable synchronization possible, wherein nervous systems affect and motivate each other, causing a 'corporeal-synchrony induced sensation of boundary loss' (Argent, 2012, p. 121). Human religious and communal ritual often involves collective emotion and motion, which produces feelings of pleasure, elation, and transcendence. Dance, music, sports, and marches involve boundary loss, transcendence of the individual through rhythym, and ecstatic expansion. Argent closes by proposing that such pleasures in merging are not only available to the human part of the horse-human pair; to give the horse its full agency (while noting that animal exploitation and domination certainly exist), we should allow for the possibility that horse-human synchrony and entrainment – what I am calling interspecies co-embodiment – works because horses actively enjoy and seek out such corporeal-affective transcendence, as well. Achieving such an intimate, exhilarating union requires the empathetic corporeal receptivity Argent describes, as well as fully engaged attention. Horses often insist on keen attention: from a 'green' untrained colt who is unpredictable and therefore dangerous, to a 'hot' high-energy mare who is easily distracted, to a refined upper-level horse who becomes frustrated with rider sloppiness that blocks his ability to be responsive. When really attending to a horse I am working with, I am fully implicated in the experience, and not with my ego. If communication breaks down, calling it a misbehavior, 'taking it personally', or exerting angry dominance are counterproductive. Furthermore, if I am really paying attention, doing so becomes nonsensical. This attentive presence has many overlaps with what is popularly theorized as 'flow,' the state in which a person gets so involved in an activity and it is so completely enjoyable that nothing else seems to matter, often resulting in optimal performance (Csikszentmihalyi, 1990). Flow requires clear and challenging goals balanced against the skills required to meet them; once achieved, flow is characterized by total concentration, a loss of self-consciousness, a merging of action and awareness such that the activity almost feels automatic, a sense of heightened control, a distortion of time, and/or a feeling that the activity is intrinsically rewarding (Swann, Keegan, Piggott, & Crust, 2012). Flow, in turn, has many overlaps with 'mindfulness' and thus evokes affective training as well as technical capacities (Kaufman, Glass, & Pineau, 2018). Yet sport riding requires achieving flow with another being. It requires being *in relationship*, and such relationality can itself achieve a flow state. Flow in sport riding, then, involves both optimal performance at the task (dancing the dressage, jumping the jumps), and optimal performance at the relationship that enables the task. To be in relationship, even for short amounts of time, can be emotionally challenging. It is difficult to assert agency while acknowledging one's vulnerability, to be both enabled by the relationship and dependent on it, to pay true attention to another – but, arguably, more straightforward with an animal than a human, which is sport riding's unique appeal. For all its challenges, being in relationship is deeply rewarding, and sport riding allows doing so in exclusively non-verbal ways. Is this reward-challenge work or play? Does the difference exist from a horse's perspective?

Conclusion

It feels somewhere between exhaustion and elation. Feeling spent, but energized. Like uneasy excesses of energy have been siphoned off, and whatever calm, grounded energetic resources were underneath all that anxiety have been freed. Of course, on some days the feeling after riding is not so transformative. Of course, frustration and discontent and breakdown are part of the pursuit, as well. But the feeling of getting it right is rejuvenating enough and frequent enough to be highly rewarding in and of itself. I have never cared much for prizes or competition, at least not in riding. The feeling of union is what motivates me. And I can find it with novice 'green' horses as well as highly schooled ones – indeed, sometimes horses new to the game are more attentive than experienced ones who may have become bored with it or dulled to subtlety by repeated bad riding. And indeed, an untrained horse's reactivity forces my attention to the immediate moment: walking delicately on the edge of explosion, the world reduces to that one connection, and I am infinitely patient. In a life full of multi-tasking, to have even a brief moment of absolute concentration can feel profound. In a life governed by words, thoughts, and symbols, to sink into sensory kinesthetic awareness can be a relief. In a life where connection with another can seem inevitably fraught or shallow, to feel unified can unleash elation.

***

Sport riding can be considered the pursuit of interspecies co-embodiment, in which two distinct beings merge in an experience of heightened immediacy and intimacy. This bridging of consciousness during the concerted pursuit of a challenging activity, fleeting and aspirational though it may be, is, on the one hand, a counterpoint to the alienated and scattered experience that can characterize modern life, and on the other hand, bypasses the reliance on the verbal that characterizes much human sociality. These experiences are, I argue, a significant part of the appeal of sport horse riding as leisure. It is an 'escape' not just from 'work' or 'social life', but from any number of unpleasant qualities attending contemporary life in wealthy capitalist societies: ennui, isolation, distraction, anxiety, loneliness, anthropocentrism and separation from 'nature'. I am not suggesting that people need to feel miserable to be attracted to sport riding, quite the contrary; being engaged in sport riding is perhaps an effective way of mitigating potential disaffection and maintaining a wellbeing of the spirit. It is both the interspecies aspect of the sport, and the sport aspect of the interspecies engagement, that enables these unique characteristics.

Of course, there are other reasons people find riding appealing that may supplement or replace the immediacy and intimacy I am discussing. Human-horse relationships can foster mutual wellbeing and companionship in a more quotidian sense (Birke, Hockenhull, & Creighton, 2010), and provide 'therapy' for stress, in an interesting medicalization of pleasure (Davis, Maurstad, & Dean, 2015). There are social relationships with other riders, with important gendered aspects. Horses can be a way of life, a lifestyle accessory, an identity or trend (Latimer & Birke, 2009). Bodily engagement with horses has specific appeal for women, who often learn fraught relationships with their own bodies (Birke & Brandt, 2009; Brandt, 2005; Dashper, 2016). This positive quality can override parents' concerns about the sport's risks to their horse-crazy daughters (Sanchez, 2016). Exploring the ways these various aspects work together to attract and retain enthusiasts, and elaborating on aspects that make the sport unappealing to people who otherwise have the time, resources, and embodied ability to pursue it as leisure, are directions for further research.

As a teenage female rider, I heard a quip that stuck with me, proclaiming that 'A girl needs something to love when she is too old for dolls and too young for boys – a horse is good'. While I certainly appreciate the grounding, strengthening effect that riding and horse companionship can have when grappling with 'femininity' during turbulent teenage years, I marvel at how neatly this quip posits the animal as in-between inanimate object and human subject. Rather, I argue that a relationship of immediacy and intimacy with a sport horse reminds us that bodily material and mindful selfhood are one and the same. The body is the source of our capacity for connection and presence, both with another and with ourselves. In many human social interactions in contemporary Western societies, bodily engagement is fraught, circumscribed, or ignored. Physical challenges in sports do not often involve profound connections with others. Sport riding combines physical challenge and connection with another being, which is an important aspect of its current cultural relevance. It offers permission to embrace the corporeal and receptive nature of selfhood in the present moment.

Disclosure statement

No potential conflict of interest was reported by the author.

ORCID

Andrea Ford (iD) http://orcid.org/0000-0003-0462-9669

References

Allen-Collinson, J. (2011). Intention and epoche in tension: Autophenomenography, bracketing and a novel approach to researching sporting embodiment. *Qualitative Research in Sport, Exercise and Health*, *3*(1), 48–62.

Argent, G. (2012). Toward a privileging of the nonverbal: Communication, corporeal synchrony and transcendence in humans and horses. In J. A. Smith & R. W. Mitchell (Eds.), *Experiencing animal minds: An anthology of animal–Human encounters* (pp. 111–128). New York: Columbia University Press.

Birke, L. (2008). Talking about horses: Control and freedom in the world of "natural horsemanship". *Society and Animals*, *16*(2), 107–126.

Birke, L., & Brandt, K. (2009). Mutual corporeality: Gender and human/horse relationships. *Women's Studies International Forum*, *32*(3), 189–197.

Birke, L., Hockenhull, J., & Creighton, E. (2010). The horse's tale: Narratives of caring for/about horses. *Society and Animals*, *18*(4), 331–347.

Brandt, K. J. (2005). *Intelligent bodies: Women's embodiment and subjectivity in the human-horse communication process* (Dissertation). University of Colorado, Sociology, ProQuest Information and Learning.

Brown, S.-E. (2007). Companion animals as selfobjects. *Anthrozoös*, *20*(4), 329–343.

Butler, D. (2017). Regaining a feel for the game through interspecies sport. *Sociology of Sport Journal*, *34*(2), 124–135.

Csikszentmihalyi, M. (1990). *Flow: The psychology of optimal experience*. New York: Harper & Row.

Csordas, T. J. (1993). Somatic modes of attention. *Cultural Anthropology*, *8*(2), 135–156.

Dashper, K. (2016). Strong, active women: (Re)doing rural femininity through equestrian sport and leisure. *Ethnography*, *17*(3), 350–368.

Dashper, K. (2017). Listening to horses: Developing attentive interspecies relationships through sport and leisure. *Society & Animals*, *25*(3), 207–224.

Davis, D. L., Maurstad, A., & Dean, S. (2015). My horse is my therapist: The medicalization of pleasure among women equestrians. *Medical Anthropology Quarterly*, *29*(3), 298–315.

Despret, V. (2004). The body we care for: Figures of anthropo-zoo-genesis. *Body & Society*, *10*, 111–134.

Ellis, C. (1999). Heartful autoethnography. *Qualitative Health Research*, *9*(5), 669–683.

Evans, G. E. (2008). *The horse in the furrow*. London: Faber and Faber.

Fureix, C., Jego, P., Sankey, C., & Hausberger, M. (2009). How horses (equus caballus) see the world: Humans as significant "objects". *Animal Cognition*, *12*(4), 643–654.

Game, A. (2001). Riding: Embodying the centaur. *Body & Society*, *7*(4), 1–12.

Grosz, E. (1994). *Volatile bodies: Toward a corporeal feminism*. Bloomington: Indiana University Press.

Grosz, E. (2005). Chapter 9: Prosthetic objects. In *Time travels: Feminism, nature, power* (pp. 145–152). Crows Nest, Australia: Allen and Unwin.

Hall, E. T. (1983). *The dance of life*. New York: Anchor.

Haraway, D. J. (2008). *When species meet*. Minneapolis: University of Minnesota Press.

Hausberger, M., Roche, H., Henry, S., & Visser, E. K. (2008). A review of the human-horse relationship. *Applied Animal Behaviour Science, 109*(1), 1–24.

Kaufman, K. A., Glass, C. R., & Pineau, T. R. (2018). Going with the flow: Mindfulness and peak performance. In *Mindful sport performance enhancement: Mental training for athletes and coaches* (pp. 47–59). Washington: American Psychological Association.

Keeling, L., Jonare, L., & Lanneborn, L. (2009). Investigating horse-human interactions: The effect of a nervous human. *Veterinary Journal, 181*(1), 70–71.

Latimer, J., & Birke, L. (2009). Natural relations: Horses, knowledge, technology. *The Sociological Review, 57*(1), 1–27.

McLean, A. N., & McGreevy, P. D. (2010). Ethical equitation: Capping the price horses pay for human glory. *Journal of Veterinary Behavior: Clinical Applications and Research, 5*(4), 203–209.

Merleau-Ponty, M. (1962). *The phenomenology of perception*. London: Routledge.

Merleau-Ponty, M. (1968). *The visible and the invisible*. Evanston: Northwestern University Press.

Merleau-Ponty, M. (2003). *Nature: Course notes from the Collège de France*. (Seigland, D., compiler. Vallier, R., trans.). Evanston: Northwestern University Press.

Oma, K. A. (2007). *Human-animal relationships: Mutual becomings in scandinavian and sicilian households, 900–500 b.c.* Oslo Arkeologiske Series No. 9. Oslo: Unipub.

Sanchez, L. (2016). 'Every time they ride, I pray:' Parents' management of daughters' horseback riding risks. *Sociology of Sport Journal, 34*, 259–269.

Sartre, J.-P. (1984). *Being and nothingness*. (Hazel E. Barnes Trans.). New York: Washington Square Press.

Seaton, B. (2013). Siding with the world: Reciprocal expressions of human and nature in an impending era of loneliness. *Emotion, Space and Society, 6*, 73–80.

Sheets-Johnston, M. (1999). *Consciousness: An Aristotelian account. The primacy of movement*. Amsterdam: John Benjamins.

Sheets-Johnstone, M. (2002). Introduction to the special topic: Epistemology and movement. *Journal of the Philosophy of Sport, 29*, 103–105.

Sklar, D. (1994). Can bodylore be brought to its senses? *Journal of American Folklore, 107*(423), 9–22.

Smart, C. (2011). Ways of knowing: Crossing species boundaries. *Methodological Innovations Online, 6*(3), 27–38.

Stone, A. R. (1994). Split subjects, not atoms; or, how i fell in love with my prosthesis. *Configurations, 2*(1), 173–190.

Swann, C., Keegan, R. J., Piggott, D., & Crust, L. (2012). A systematic review of the experience, occurrence, and controllability of flow states in elite sport. *Psychology of Sport and Exercise, 13*, 807–819.

Wipper, A. (2000). The partnership: The horse-rider relationship in eventing. *Symbolic Interaction, 23*(1), 47–70.

Relating to reptiles: an autoethnographic account of animal–leisure relationships

Kevin Markwell

ABSTRACT

In this paper, I explore my life-long interest in amateur herpetology using an autoethnographic methodology and, in doing so, reveal the varied intersections of animals, leisure, place and identity within my own life experience. Informed by serious leisure theory and Stebbins' concept of the 'leisure career', I reflect on aspects of my life-history from childhood and adolescence through to the present day as a 56-year-old Anglo-Australian male, teasing out the diversity of ways that the leisure practices relating to my interest in this specific group of animals have been enabled and supported, spatially, structurally and socially. This analysis examines the intellectual, emotional and embodied ways that I have encountered and experienced reptiles across time and space, facilitated in the first instance by a supportive family which, in various ways, encouraged my entry into amateur herpetology.

Introduction

The early morning warmth of the sun seeped into my body as I lay on my grandparents' front veranda, face resting on crossed arms, peering over the edge, waiting, a little impatiently, for the creature to emerge. Sensing the warming sunshine on the grass in front of it, the blue-tongued lizard, with its vinyl-shiny, silvery-grey body, edged its way slowly into the patch of sunshine where it stopped, warming its head. I lay quietly, enthralled, hardly daring to breathe in case I scared it back under.

I was 10 years old and this was the first time that I had seen this kind of lizard in 'real-life', so close I could have reached down and touched it. The black and white photographs in books had come to life, just inches from my face. As I lay watching, fascinated, its cobalt-blue fleshy tongue poked slowly out of its mouth and it crawled further into the sunshine, revealing its 50 cm long body. I was lost in the colours and textures of its scaly body, the pin-prick-like marks that studded some of its larger head scales, the dark mask across its eyes, and repulsed by the swollen paralysis ticks that hung obscenely from its ear-holes.

I returned often during that summer to lie upon the splintery wooden boards of that veranda, watching 'my' lizard, and, from time to time, dropping banana on to the grass for it to eat. Eventually it moved on, not to be seen again. But during those moments, I was lost in time, completely absorbed in the moment, as I lay there, enjoying my close proximity to this wild animal.

I have been interested in reptiles for as long as I can remember. As a young boy I caught small lizards and placed them in empty cans so I could closely watch them scurrying about, observing their gulping throats as they breathed, before releasing them into the garden. Growing up in the

1970s in the working class suburbs of an Australian industrial city provided me with an array of opportunities to nourish my passion: a variety of reptiles could be found within remnant scraps of bushland, beside creeks and in the backyards of houses. As an adolescent I planned and led day-long forays and, as I grew older, camping trips, into bushland with friends, who, while not sharing my passion, were nevertheless happy to accompany me on these little adventures. I read, (and continue to read), voraciously about reptiles and, in doing so I have developed literacy in reptile natural history that goes beyond knowledge acquisition and into an understanding of how that knowledge was created. I took (and take) advantage of any opportunity to further my connections with reptiles that was presented to me.

Although the expression of my own interest in reptiles has varied across my life course, it has been a constant and personally significant dimension of my identity constituted predominantly through leisure, which I conceptualise here as a mix of discretionary times and accessible spaces through which I engage intellectually, corporeally, imaginatively and creatively with reptiles. As a group of animals, reptiles are much less popular and 'loved' by humans than are birds, mammals or even fish or butterflies. Indeed the Australian reptile authority and someone who has been an important influence on my interest in reptiles, Eric Worrell, wrote of them as the 'unpopular ones' (Worrell, 1966, p. 90). Snakes, in particular, sliding about on limbless bodies, with their unblinking stare, an ability to shed their skins intact and swallow their prey whole, elicit a variety of extreme reactions and attitudes: revered and worshipped, or regarded as evil, malevolent beings and feared (Morris & Morris, 1965; Nissenson & Jonas, 1995; Stutesman, 2005).

Reptiles have received little academic attention as objects of leisure, and I was unable to locate any studies that have examined interest in reptiles as a focus of leisure experience. Instead, publications cover the ecological impacts of harvesting of reptiles for the pet trade (see for example Natusch & Lyons, 2012), infections arising from pet reptiles (Embil & Nicolle, 1997), impacts of recreational activities on reptile populations (Bowen ad Janzen, 2008) or as foci of wildlife tourism, such as marine turtles (Tisdell & Wilson, 2001) and crocodiles (Ryan & Harvey, 2000). Although there are no data on the popularity of amateur herpetology, the growth in popularity of reptiles as pets has been reported on (Koehler, 2017) and a search on Facebook (13 October, 2018) reveals an array of reptile-related sites including Australian Reptile Enthusiasts (12,575 members), Reptile Connection (35,000 members), Reptiles Passion (27,000 members) and *In Situ* Field Herping Photography (3,222 members).

Theorising animal-leisure relationships

Bulliet (2005) has argued that contemporary Western engagements with non-human animals occur within what he has called the 'post-domestic' phase of the West's relations with animals. Domesticity, in which our dominant relations with animals were framed within farming, has, he argues, given way to post-domesticity, characterised by a spatial and seemingly emotional detachment from the animals that provide us with meat, milk and leather. Contemporary relations with animals in the West, Bulliet suggests, are estranged from those animals that produce these commodities. Additionally, we no longer construct the greater majority of animals as competitors for resources such as food or threats to our individual or collective well-being.

Following on from this argument, for the greater majority of us living in Western societies, our lived experiences of non-human animals tend to take place, predominantly, within the broad array of leisure practices, structures and spaces. Yet, the human-animal studies literature appears not to have recognised, at least explicitly, the importance of leisure as a domain through which human–animal relationships are contextualised. The word 'leisure,' for instance, does not appear in the index of 10 monographs on human–animal studies that I examined, although 'entertainment' was listed in three of them.

Domesticated or wild, non-human animals are our pets or companions; the subjects of all manner of popular culture: books, magazines, documentaries, films, websites, social media and

YouTube videos; a key driver for many travel experiences; and the targets of those who engage in the leisure practices of photography, hunting and fishing. While the majority of research into leisure-animal relations has involved domesticated animals, there is a small but growing body of work on the ways in which wild animals are entangled with human leisure that has been undertaken by scholars in leisure studies (see for example Carr & Young, 2018). As Ferguson and Litchfield (2018, p. 12) observe, 'there are now countless opportunities for people to engage with wild animals as part of their leisure time'. The ways we make meaning of these animals are shaped and facilitated by leisure practices and structures, which are themselves contextualised within particular histories, geographies, economies, societies and, increasingly, technologies.

The rapid growth of human–animal studies over the past decade reflects the increasing preoccupation we have with understanding critically our relationships with animals. The multidisciplinary field of study attempts to interrogate and understand the ways we make meanings of our encounters with non-human animals, revealing the contradictions, inconsistencies and complexities of those encounters (Demello, 2010; Herzog, 2010; Waldau, 2013). The 'animal turn' in leisure and tourism studies, although somewhat belated, is reflected in a growing body of publications including number of monographs and edited collections that have been published in the last few years (see for example Carr, 2014, 2015; Carr & Young, 2018; Dashper, 2017; Markwell, 2015), and, indeed, by this special issue.

In attempting to make theoretical sense of my lived experiences of the ways by which my involvement with reptiles has constituted so much of my leisure time and experience, I am drawn to Stebbins' concept of serious leisure. Serious leisure is '…the systematic pursuit of an amateur, hobbyist or volunteer activity that people find so substantial, interesting and fulfilling…that they launch themselves in a (leisure) career centred on acquiring and expressing a combination of its special skills, knowledge and experience' (Stebbins, 1992, p. 3). In his continued development and refinement of the concept expressed in an extensive series of publications, Stebbins has identified six qualities that characterise serious leisure pursuits: construction of a leisure career; need for occasional perseverance in the face of difficulty; application of effort and the acquisition and application of knowledge and skills; durable benefits such as self-actualisation, self-enrichment, and enhanced self-image; a sense of belonging to the social world associated with the pursuit; and a strong sense of identity with the pursuit (Stebbins, 2007, pp. 11–13). Involvement in serious leisure will necessarily structure the individual's use of their free time and impose certain costs (time, financial, and, in some cases, personal).

Finding my stories, telling my stories

I use autoethnography as a method of critically reflecting on my own life experiences to uncover and examine the intersections between reptiles, leisure and identity that have occurred within my own life, contextualised within the Australian cultural milieu through which I am constituted and concomitantly, help constitute. Autoethnography is a qualitative method that allows subjective interrogation of the ongoing connections and reconnections between self and context (Ngunjiri, Hernandez, & Chang, 2010). Anderson and Austin (2012) argue that autoethnography is 'well-suited' to leisure studies research as it enables a nuanced, often deeply personal, subjective and embodied account and analysis of leisure practices under study. Rather than writing the researcher out of research, autoethnography requires an active, authorial voice to take front-stage. These authors recognise two forms of ethnography: analytical and evocative. The approach I take in this article is predominantly analytical in that I attempt to understand and explain my relations with reptiles conceptually through drawing on the theory of serious leisure and the ideas and arguments of human–animal studies. However, I hope that the accounts of my leisure involvements with reptiles are engaging and evocative and elicit emotional responses from the reader such that the reader 'feel[s] the feelings of the other' (Denzin, 1997, p. 228).

Utilising an autoethnographic approach, and informed by a serious leisure theoretical perspective, it is my intention in this paper to critically interrogate my personal relationships with one class of animals, *Reptilia*, so as to make a contribution to better understanding the interactions and relationships between human and non-human animals that are created and constructed, nurtured and sustained, disrupted and disturbed by leisure practices, spaces and contexts. I aim to reveal and critically interrogate the intersections between reptiles, leisure, place, space and time.

What are these autoethnograpic data that I have drawn upon to make sense of my ongoing fascination for reptiles that has been sustained across time and space, for almost five decades? There is memory, of course, and I have utilised the store of strong memories that I have added to over this time. Some of these, like my first encounter with a blue tongued lizard that began the article, are vivid and durable. Others have been 're-remembered' through immersion in the personal archive that I have inadvertently created from my habit of keeping so much of the flotsam and jetsam of my life, the artefacts of my life that have survived multiple house moves and episodic attempts to de-clutter my life.

DeLyser, 2014, p. 209) has coined the term 'archival autoethnography' to describe the practice of personally collecting and keeping an archive of memorabilia, artefacts and other objects of a life with which to engage critically. Following DeLeyser, I have used my own personal archive to help me (re)connect to people, animals, places and events over my life that have been entwined with my interest in reptiles. My archive includes photographs, loose, in albums and in digital form; many boxes of photographic slides; letters I have received and drafts of letters I have written to other amateur and professional herpetologists; scrapbooks of newspaper clippings dating back to when I was 10; notebooks that I have kept in at various stages across my life; birthday cards that feature depictions of reptiles; references written by school principals, academics and employers; my published (non-academic) writings such as letters to newspapers, and short articles for newspapers and magazines; and an assortment of other material that has survived more or less intact. The use of these other materials allows me to go beyond a reliance on memory as my only source of data.

Over the past 2 months that I have been working on this paper I have immersed myself in this material, reminiscing on long-forgotten aspects of my life. I have discovered things that I had written or activities in which I had been involved which I had completely forgotten about, yet filled me with some pride once discovered. While some authors have noted that doing autoethnography can be a problematic enterprise because it can uncover and bring to the surface stories of 'loss, pain, grief, depression, eating disorders, family drama' (Ngunjiri et al., 2010; np; see also Dashper, 2017) my experience has been one of rediscovery of (mostly) enjoyable, joyful and interesting events and individuals. Remembering, rediscovering and writing have been positive and enjoyable processes. I have also used my library of reptile books as a means to better understand my relationships with reptiles through leisure. As I discuss later in the paper, books have played a big part of my biography since childhood. Imposing some analysis on to my collection has enabled me to get a sense of the temporal patterns of my book buying, the authors who I value most, and the books that I keep revisiting.

The greatest methodological challenge that has created much angst has been to decide on what I could focus on within the word limitations imposed by the article. The first draft was a document almost 20,000 words as I wrote and re-wrote of the reptile-related constituents of my life. Initially I was intending to sharpen my focus on three, then two, aspects of my leisure interest that I thought could serve to provide both empirical and theoretical anchors to the narrative. But as I continued with the process of writing and rewriting this approach still seemed artificial and derivative. On a break from writing, I walked up to the large outdoor enclosure in which I keep a breeding colony of blotched blue tongued lizards. One was sloughing its skin in patches, revealing the fresh, colourful skin underneath and it struck me that this is what I was doing in this paper: sloughing a series of patches or vignettes from my life through which an underlay of meanings could be revealed and, hopefully, understood. These patches, some short, some longer, and, admittedly, simply fragments of my life, provide me with the opportunities to tease out and reflect on my relations with reptiles framed by, and constituted within, leisure.

Reptiles as undesirable wildlife: an early awakening of difference

I'm five or six years old. I stand with my mother and younger sister at the back of our rented house, look across the patchy lawn, past the Hills Hoist clothes line, towards the back fence of wooden palings that separate our place from the bushland that grew alongside Dead Dog Creek, a creek which would, in a few years, be drained and lined with concrete. My father, with the wooden handle of the garden hoe held high above his shoulders, runs fast after that streaky blur of black and red, the snake holding its head off the ground fleeing for its life, ultimately escaping through a hole in the fence and into the scrub. I told Mum that I was glad that Dad hadn't killed the snake.

The Australia I grew up in was one in which 'the only good snake was a dead snake'. Snakes were regarded by Australian culture as vermin, as undesirable animals that were out of place in suburban back yards. Such 'out-of-place-ness' meant that snakes, and often lizards as well, were killed without hesitation. Indeed, the bodies of snakes killed with a shovel would sometimes be displayed on local garden fences (I would take these and preserve them in glass jars filled with methylated spirits that would be displayed on my bookshelf in my bedroom). At 9, I was devastated when a neighbour showed me the newly killed body of a beautiful green tree snake, which had 'crossed the line' and entered his back yard from neighbouring bushland. He had killed (the completely harmless) snake with a horse brush. I soon understood that reptiles were undesirable and that my interest in, and passion for them, was unusual and a bit quirky.

As Markwell and Cushing (2010, p. xiv) have observed, 'A deep-seated anxiety about snakes was carried to Australia, stowed in the cultural baggage of British colonists, and passed down for generations'. Lizards generally fared better, although these too were frequently killed. The bushland remnants and creeks that still persisted in the suburbs and the incursions of housing developments into bushland on the peri-urban fringe brought reptiles and suburbanites into frequent contact, often to the detriment of the 'invading' reptile. The geography of human–animal relationships is central in understanding the contours, real or imagined, that map out socially appropriate interactions between human and non-human (Urbanik, 2012). Looking at and reflecting on my photographs, my notebook scrawling, and my published writing, I can see now that my leisure practices have consciously sought to transcend those contours and break down the barriers between reptile and human by enabling reptiles to inhabit or co-inhabit the material and imaginative spaces that I have occupied.

The separated-ness of reptiles from other animals was reinforced in law. Unlike native bird and mammal species, no legislation existed to protect reptiles until 1974 when the National Parks and Wildlife Act was amended to include reptiles (Lunney & Ayers, 1993). Up until then, it was legal to kill or catch and keep reptiles. In this sense, reptiles were treated by the state as inferior, 'other' animals, not worthy of protection. This pariah status continued to shape social attitudes towards reptiles, and snakes, in particular.

However, the fact that reptiles weren't protected also meant that, for kids like me, local reptiles could be caught and kept as pets, enabling me to facilitate the incursion of reptiles into the domestic realm of my family, mostly by keeping them in tanks and outdoor enclosures that became significant components of the domestic semiotics of my childhood house and backyard (Rainbird & Roswell, 2011). The remnants of habitat also meant that I had access to a diversity of reptiles which I could find, watch and catch. Interacting, either directly or indirectly, with reptiles became, from the age of about 10, my primary leisure practice right through my adolescence. Numerous photographs, some black and white, some colour, depict me with pet lizards, snakes that I had caught, and a variety of reptiles that I had observed in the wild, often with accompanying habitat photographs. Birthday cards with illustrations of snakes, lizards and turtles, given to me by family or friends, indicate that even by early adolescence the importance of reptiles in my life, and to my sense of self, was apparent to others.

Captured within books: reading about reptiles

Reading about reptiles was, and still is, for me, a solitary pleasure. Growing up as a boy in the Australia of the late 1960s and 1970s was predominantly an outdoor enterprise involving organised competitive team sports: cricket in the summer and some kind of football in the winter. My parents quietly accepted that I wasn't a sporty kid and did not push me into sport, although I endured a season playing cricket as a 13 year old, which I hated. I've never thought of myself as bookish, which might seem a little odd given that across from the desk on which I am working now are bookshelves covering an entire wall of my study, which house my collection of nearly 2000 books, of which about 300 are reptile-related. Nevertheless, I am a reader. I have always been and I suspect I always will be. I bought my first book with money given to me by my mother when I was 9 years old: *The King Bear* by Michael Turner (published in 1968). I still have this book. Reading has helped build my knowledge in all kinds of areas; inspired me; comforted me; motivated me; and propelled me into imaginative worlds, reptilian and otherwise.

Through my appetite for reading about reptiles, I developed knowledge about these animals by the time I was a mid-teen that was valued and acknowledged by my parents and other relatives. Reading enabled me to develop confidence in my knowledge of reptiles which I would share through short talks during school assemblies and, later, more sophisticated lectures to all manners of community groups and as part of environmental education programs in national parks. The knowledge that I developed, largely through reading, and the authority on which I could talk to others about reptiles, became an integral part of my identity as an adolescent and young man. I have no doubt it marked me off as a little different from others, but I also think it helped deflect attention away from some of my 'flawed masculinity' when it came to lack of interest in sport, motorbikes and indeed, girls. Among my peer group, being able to catch snakes was a reasonable, masculine substitute, for playing footy.

The leisure studies literature is remarkably quiet on reading as leisure, with most literature on the topic contributed by education and literacy disciplines. In a recent Australian study into reading for pleasure, Throsby, Zwar, and Morgan (2017) found that 65% of respondents read at least once per week, and that reading for pleasure was the third most popular leisure activity after browsing the Internet and watching TV. About 40% of the samples were categorised as 'frequent readers' (those who read more than 10 books in a year). Only about a third of frequent readers were males, and males constituted the highest category of non-readers. When asked why they read for pleasure, three broad categories were identified by the authors: enjoyment, learning and health and wellbeing (Throsby et al., 2017, p. 9)

While several authors have identified the role that a family culture of reading and of valuing books and literature has on children's reading-as-leisure habits (see Rainbird and Roswell, 2011; Hughes-Hassell and Rodge, 2007), there was no culture of reading in my working-class family, and there was not even a bookshelf in the living room of my childhood home. My father had little formal education and was illiterate up until his mid-50s. My mother, like many of her contemporaries, read Mills and Boon romances, Yates' Garden Guides and women's magazines. However, both parents encouraged me to read: 'That's the way you learn things, Kev. You've got to read books'. And read, I did. I soaked up information about reptiles, I covered distribution maps with grease-proof paper so I could trace over them and commit these distributions to memory, and, following my grandfather's advice, I learned to memorise the scientific names of the species about which I was reading. There are pages and pages in my childhood notebooks of scientific names written and re-written to rote learn these often difficult names.

My school library and local public library became significant leisure spaces. Although my relatives soon realised that books were the ideal birthday and Christmas gift for me, the bulk of my reading up until my mid-teens involved borrowing library books and spending time sitting in libraries reading those magazines and journals on animals I was unable to borrow. I was so much more confident in the library space than on a sporting field and the quietness and lack of competitive or combative attitudes I found calming. I understood the Dewey decimal system, call numbers and the cabinets of reference cards, and by the age of 10 I was requesting inter-library loans through my public library. The school library, in particular, I remember as a safe space, calm, organised, free of the hyper-masculine hijinks and bullying of the playground outside. It was a sanctuary that also gave me access to desired information about reptiles (and other wildlife).

While the range of book titles on reptiles was nowhere near as extensive as it is today, I nevertheless borrowed from my local library a number of books on Australian reptiles written by authors Eric Worrell, Harry Frauca, David Fleay and Hal Cogger. Eric Worrell in particular, was the primary authority on Australian reptiles as I was growing up and he soon became a childhood hero of mine. I wanted to be Eric Worrell! Worrell was a self-made naturalist/herpetologist, who had left school at around 13 but who, at the age of 39, published *Reptiles of Australia*, the first authoritative, comprehensive text on Australia's reptile fauna. This was a scholarly, scientific, taxonomic work that included ecological information that had never been published before along with photographs, many in colour, of species which had never been photographed. For anyone interested in Australian herpetology, this book was a game changer.

As my ability to buy books grew, I began building a collection of books on reptiles that today numbers almost 300. I scour catalogues of newly published books and regularly visit blogs and websites that provide me with information about new books about reptiles. I am a devoted explorer of second-hand bookshops on my quest for rare and unusual reptile (and other wildlife) books and I regularly visit eBay and other online sites looking for books. I am no longer just a reader of reptile books, but a dedicated collector, willing to pay irrationally large amounts of money for particular titles or first editions. Searching for these books involves the commitment of resources: time and money, as well as utilising my knowledge of publications on reptiles.

Australian snake man: Eric Worrell

Worrell took on heroic status for me during my childhood and adolescence. Just as other boys idolised rugby league and cricket players, I idolised him. In my personal archive there are all kinds of 'Worrellian' paraphernalia: newspaper articles pasted into scrap books; magazine articles; guidebooks from his Australian Reptile Park; and on my bookshelves are all his published books. Through his books and articles, I learned how to be an 'amateur herpetologist': memorising the knowledge that he had shared through his writings; working out how to go about finding reptiles in the wild and mounting expeditions to find them; and reading about how to care for them in captivity. The fact that he had begun his interest just as I had: keeping local reptiles in cages and enclosures built by his father and leading little expeditions into local bushland in search of them made him even more accessible to me.

But it was visiting his Reptile Park at Wyoming, 80 km from the house I grew up in, that had the most profound, most visceral impact on me as a boy and young man:

As we pulled off the highway into the carpark out the front, protected by the life size replica of a Diplodocus dinosaur, my belly filled with butterflies. Once we had our entry tickets, I passed through a threshold into another world, one in which reptiles were displayed, admired and celebrated. Local species of reptiles were exhibited in large open cement walled, 'pits'. What appeared to be dozens of red bellied blacks, tigersnakes and brownsnakes sunned themselves in tangled piles a metre or so from my body. I leaned against the warm concrete-rendered wall, transfixed and in sublime wonder, watching them flick out their tongues, slide across each other's body, or investigating crevices in their enclosure.

Further into the heart of the Park, I stand in quiet awe at the world's largest venomous snake, the king cobra of south-east Asia. Two individuals, a 'king' and 'queen' inhabit this large exhibit, each between four and five metres long. I have admired them on previous trips, but today they are even more impressive, more compelling: they are both active, sliding across the floor of their enclosure, poking their fist-sized blunt heads into crevices. At times they seem to look at me. I am completely captured by their presence.

Keeping reptiles

*Stony-faced, mum handed me the newspaper article that my nanna had cut out from the Sunday paper soon after I had arrived home from school. 'SNAKES ALIVE! YOU CAN'T WIN' screamed the headline in bold, block letters. Underneath, the subheading pierced my brain: **Ban on keeping reptiles**. 'Attention, reptile owners. New protection laws now prohibit the keeping of most species of reptiles.' I couldn't believe it. Keeping reptiles was **who I was**. I took the article and headed into my room, closed the door, and began to sob. I read and re-read the article, trying to make sense of it, trying to understand what this meant for me. Although the new legislation identified a number of species that could be kept legally in quantities of two individuals, other species (such as a trio of bearded dragons, named Cuthbert, Horace and Cordelia) that I was keeping at the time could no longer be kept without being registered as a 'reptile fancier'. The article ended 'Registration will be granted only if the [National Parks and Wildlife] Service is satisfied that the reptiles are being kept correctly'.*

When, in 1975, at the age of 13, legislation was brought in that protected reptiles and made it illegal to collect them from the wild and illegal to keep them without a licence, I was shattered. It felt like the thing that made me who I was, was being ripped away from me. My identity, even at the age of 13, had been largely based on my involvements with reptiles, more particularly, through the acquisition of knowledge through avid reading and through catching and keeping these animals. My initial fears, however, that I would no longer be able to keep my reptiles were unfounded as I could, and did, apply for a licence which enabled me to keep most of my reptiles, legally, although it was no longer legal to collect reptiles from the wild. Nevertheless, my hobby had begun to be constructed as 'leisure on the margins of conventional morality' (Lynch & Veal, 2006), and people who were found to be keeping reptiles illegally were subjected to significant fines and public humiliation.

I ended my keeping of reptiles sometime in the mid-1980s once the animals that I had been keeping at my parents' place since I was a boy had died of old-age. However, in 2006, at the age of 44, I began keeping reptiles again. Changes to legislation had occurred early in the 2000s which created a three-tiered licencing structure covering the greater majority of Australian species. In addition, the Internet had now become a vast marketplace for the selling of Australian reptiles with a variety of websites and Facebook pages through which reptiles were bought and sold. For instance, as an example, Australian Reptile Sales, a Facebook Page https://www.facebook.com/groups/531184326936721/, had 18, 219 members as of 26 September 2018. Within NSW legislation, private keepers only have access to captive-bred animals. At the time of writing this article I keep a woma python called Kimba, a pygmy monitor and a breeding colony of around a dozen blotched blue tongues.

Going to Bali to visit the king: February 2016

The heavy rain the day before had left the ground sodden and muddy as the six of us (three reptile-fancying tourists: me, a Swiss fellow and a young guy from UK as well as three members of the Bali Reptile Rescue team) squelched our way through tall bamboo thicket growing alongside a swiftly flowing milky stream in the west of Bali. We were in search of the longest and arguably, the most infamous of all venomous snakes, the king cobra. We jumped down on to the rocky bed of a smaller tributary and a couple of the BRR team began poking hooks into holes and under rocky ledges. I was somewhat surprised and uncomfortable about this level of intervention as I had assumed that any cobras we found would be out in the open and not hiding in a refuge somewhere. Finding nothing we continued up and out of the cooler gully, walking up a spur to a ridge in hot sun and through scrubby, savannah like vegetation.

Later, at lunch, a call came that a king cobra was near a farmhouse a few minutes away. We jumped in the jeep and sped off. I was still not convinced that I would see a real live king cobra as we trudged past smelly pig sties and little kids playing in the mud between wicker baskets containing fighting cocks. And then I turned a corner and there, less than two metres in front of me, was a three metre long king cobra, indignant and cranky, its fore-body raised almost a metre above the ground, its hood spread, its jaws open, revealing a pink mouth and fleshy fang sheaths. It made a sound completely unlike any other I had heard a snake make – more a growl than a hiss. I stood completely in awe of this extraordinary snake with a body thicker than my arm and venom that could kill an elephant.

The cobra was fairly easily caught – king cobras stand their ground, but they are relatively slow compared to most other snakes. Fixated on the left hand of the cobra wrangler, the snake was unaware of the other, gloved, hand that was slowly sliding up what might be considered the throat of the snake until he closed quickly but gently around the great snake's neck. After some photography, the cobra was placed in a bag, and taken back to the BRR depot, where it would wait until being re-located somewhere where it didn't run the risk of being blasted with a shot gun. I felt exhilarated, still high on the adrenalin rush that surged through my body when I first saw the snake, which was so indignant, so powerful. I was happy that the snake had been rescued, it would have surely been killed had we not arrived when we did, but I also know that relocating snakes is not straightforward, with radio telemetry studies showing that relocated king cobras move about far more and cover distances much greater than resident cobras, putting themselves in dangerous situations when they find themselves close to human activity (Barve, Bhaisare, Giri, Gowri Shankar, Whittaker and Goode, 2013).

Later that night we trudged along narrow paths following irrigation channels through rice padi looking for the reptiles that still manage to exist in such a human-modified landscape. This was a reptile-lover's paradise: skinny bronzeback snakes draped on the branches of shrubs; banded krait; spitting cobra; reticulated python; water monitor…species I had never seen in the wild, all here, in this rice padi on an island best known as a mass tourism destination.

My reptilian leisure career

Over the course of my life my passion for reptiles has been experienced and expressed in a variety of ways: cognitively, affectively and corporeally. Through the practices, structures and spaces that constitute my leisure I have constructed what Stebbins' (2007) has termed a leisure career, through which individuals build a corpus of specialised knowledge, skills and experience that provides a sense of fulfilment, satisfaction and purpose.

I recognise, within the way I have constructed my leisure-interest, a form of scholarship that has enabled me to develop considerable knowledge and skills, (a 'literacy' in reptilian natural history), and, as a consequence, the ability to 'talk the talk' with other people interested in herpetology, both amateur and professional. The development of deep knowledge, not just based on book-learning but through original observations, appears to be common amongst many wildlife-based 'hobbyists' whatever their chosen group of animals. These individuals develop caches of knowledge and wisdom, skills and expertise, which often parallel their professional counterparts (see for example studies of amateur birdwatchers (Kellert, 1985; Kim, Scott, & Crompton, 1997; Connell, 2004) and insect enthusiasts (Lemelin, 2013)). These studies show that animal-based enthusiasts (and here we could also include hunters and fishers, people who keep and breed aquarium fish, or keep and show dogs or poultry) devote considerable resources in terms of time, money and intellectual and physical energy, make various sacrifices in terms of other missed opportunities, and constitute what Tomlinson (1993), in his critical contemplations of Stebbins' serious leisure concept, has called a 'culture of commitment'. Within the specific context of amateur herpetologists, Professor Rick Shine, one of the world's leading reptile ecologists, has observed that hobbyists have played a significant role in contributing to understanding the natural history of Australian reptiles (Shine, 1991).

While I had little self-awareness of the overall importance of my passion as a younger person, it is when I have critically reflected on the broad sweep of my life through this autoethnography that I can recognise a sustained, systematic, and to varying extents, disciplined commitment to the pursuit of knowledge about reptiles and seeking to have experiences with reptiles. Through these cognitive, affective and embodied experiences I have developed a strong sense of empathy with them. Leisure, constituted through discretionary time and accessible spaces, has provided the opportunities to construct what has become a highly rewarding and engaging life-long leisure career (Stebbins, 1992, 2007). My reptile-related leisure practices: reading, writing, presenting talks, photography, preserving their dead bodies, keeping and breeding them, travelling locally and internationally to see, observe and in some cases studying them in the wild and in captivity, trawling the Internet for websites, hunting and collecting rare books about them, engaging with others, and buying reptilian-inspired art, have constructed individual and social identities that I have maintained, refined and redeveloped over the course of my life. While the acquisition of knowledge has been central, my leisure career, my serious leisure, has stimulated my imagination and creativity that have opened all kinds of enjoyable experiences.

Invoking Stebbins' characteristics of serous leisure (Stebbins, 1992, 2007), over the almost five decades that I have been interested in amateur herpetology I have developed knowledge through scholarship and sophisticated skills such as correctly identifying species using technical taxonomic keys. I have committed considerable time, energy, perseverance and money to my passion and in return I have been rewarded with a growing sense of competency and proficiency, which in turn has enriched my sense of self and the construction of an individual and social identity. I have a sense of belonging to a social world and to an underpinning ethos or set of beliefs relating to the study and enjoyment of reptiles.

Significantly, the trajectory that I have followed in developing this leisure career in amateur herpetology has been one that has been largely self-directed: leisure has given me the freedom to develop my interest in my particular way, as opposed to a professional who would have had to have taken a more structured, proscribed route. Such freedoms, the freedom to do, the freedom to have, the freedom to be, are fundamental to contemporary conceptions of leisure and of the enjoyment and pleasure that are created through it (See Kelly, 1982; Roberts, 2006). While Carr (2017, p. 148) critiques notions of freedom in leisure and suggests that much contemporary leisure is 'consumerist recreation', which lacks a sense of freedom that is truly liberating and creative, I suggest that within the concept of serious leisure, freedoms to be and to become, exist, enabling proponents to engage creatively, intellectually and corporeally with their subject.

This is not to suggest that I naively accept that my leisure career has been underpinned by unrestricted freedoms, or that individuals are not constrained by social structures such as gender, ethnicity, social class, age and sexuality. Indeed, I am acutely aware of my social situation and good fortune that allows me the income and time to pursue my interest: to purchase expensive books and photographic equipment and to travel overseas, for example. This autoethnographic project has also revealed for me the significant enabling influences of my parents, who indulged my interest, allowing me to keep reptiles at home, taking me to wildlife parks, buying me books and encouraging me to read. In this way, family becomes an important enabler of leisure (Roberts, 2006; Zabriskie & McCormick, 2001) and the subsequent development of serious leisure careers. I am also aware of the particular social and cultural geographies that shaped my early entry into amateur herpetology: the financial constraints imposed by growing up in a working class family; the opportunity constraints and lack of social capital growing up in a regional city without a major natural history museum or easy access to others interested in herpetology.

I began this paper with a memory of an encounter with a blue tongued lizard, and I want to bring this paper to an end, recounting another. On Saturday 11 February, 2006, I had a Letter to the Editor, *The Newcastle Herald*, published under the title 'Give bluey a bit of space'. It read, in part, '*What this means is recognising that bluetongues still live and breed in inner-city suburbs...They don't need rescuing and relocating. What they do need is more habitat where they have a chance of keeping out of the way of predatory dogs and cats. Make your gardens more bluey-friendly so that your kids and grandkids can still admire this beautiful animal in years to come*'. Throughout my reptilian leisure career, there have been opportunities to stretch beyond my own personal interests and pleasures, to actively engage with conservation issues, both specifically related to reptiles, such as the loss of safe habitat for urban blue-tongued lizards and those issues relating to environmental conservation and sustainability, more broadly. The empathy I have developed for reptiles and the ecological understandings of them and their role in ecosystems propel me to advocate for their interests wherever and whenever I can. The constellation of leisure practices, structures and spaces, have, and continue to shape, my relationships with reptiles, and more broadly, other wildlife and ecological systems. Leisure has been the portal through which my relationships with what I regard as an entirely fascinating and enthralling group of animals, have been ignited, developed and nourished.

My form of serious leisure, and the subsequent leisure career that has developed, however, without the structures or systematic training or facilities provided by government or community organisations for children's sport or by the scouting movement. Instead it was created and nourished through parents who were happy to support my unusual interest in various ways; opportunities for embodied encounters with all manner of reptiles in close proximity to my neighbourhood; significant individuals who, through their writings or television appearances become mentors or role models for me; and the information from reading books that I was able to transform into knowledge.

Disclosure statement

No potential conflict of interest was reported by the author.

References

Anderson, L., & Austin, M. (2012). Auto-ethnography in leisure studies. *Leisure Studies, 32*, 131-146. doi:10.1080/02614367.2011.599069

Barve, S., Bhaisare, D., Giri, A., Gowri Shankar, P., Whitaker, R., & Goode, M. (2013). A preliminary study of "rescued" King Cobras (*Ophiophagus hannah*). *Hamadryad, 36*, 80–86.

Bowen, K. D., & Janzen, F. J. (2008). Human recreation and the nesting ecology of a freshwater turtle (*Chrysemys picta*). *Chelonian Conservation and Biology, 7*, 95–100.

Bulliet, R. W. (2005). *Hunters, herders and hamburgers: The past and future of human-animal relationships.* New York: Columbia University Press.

Carr, N. (2014). *Dogs in the leisure experience.* Wallingford: CABI.

Carr, N. (Ed.). (2015). *Domestic animals and leisure.* Basingstoke: Palgrave MacMillan.

Carr, N. (2017). Re-thinking the relation between leisure and freedom. *Annals of Leisure Research, 20*, 137–151.

Carr, N., & Young, J. (Eds.). (2018). *Wild animals and leisure: Rights and wellbeing.* London: Routledge.

Connell, J. (2004). Birdwatching, twitching and tourism: Towards an Australian perspective. *Australian Geographer, 40*, 203–217.

Dashper, K. (2017). *Human-animal relationships in equestrian sport and leisure.* Abingdon: Routledge.

DeLyser, D. (2014). Collecting, kitsch and the intimate geographies of social memory: A story of archival autoethnography. *Transactions of the Institute of British Geographers, 40*, 209–222.

Demello, M. (Ed.). (2010). *Teaching the animal: Human-animal studies across the disciplines.* New York: Lantern Books.

Denzin, N. (1997). *Interpretive ethnography.* London: Sage.

Embil, J. M., & Nicolle, L. E. (1997). Salmonella urinary tract infections associated with exposure to pet iguanas. *Clinical Infectious Diseases, 25*, 172.

Ferguson, M., & Litchfield, C. (2018). Human-wild animal leisure experiences: The good, the bad and the ugly. In N. Carr & J. Young (Eds.), *Wild animals and leisure: Rights and wellbeing* (pp. 12–38). Oxon: Routledge.

Herzog, A. (2010). *Some we love, some we hate, some we eat: Why it's so hard to think straight about animals.* New York: HarperCollins.

Hughes-Hassell, S., & Rodge, P. (2007). The leisure reading habits of urban adolescents. *Journal of Adolescent & Adult Literacy, 51,* 22-33. doi:10.1598/JAAL.51.1.3

Kellert, S. R. (1985). Birdwatching in American society. *Leisure Studies, 7,* 343–360.

Kelly, J. (1982). *Leisure.* Englebert Cliffs, New Jersey: Prentice-Hall.

Kim, -S.-S., Scott, D., & Crompton, S. L. (1997). An exploration of the relationship among social psychological involvement, behavioural involvement, commitment and future intentions in the context of birdwatching. *Journal Leisure Research, 29,* 320–343.

Koehler, D. (2017). The growing popularity of cold blooded pets. *The Chronicle.* Retrieved October 13, 2018, from https://chronicle.durhamcollege.ca/2017/01/growing-popularity-pet-reptiles.

Lemelin, R. H. (2013). To bee or not to bee: Whether 'tis nobler to revere or revile those six-legged creatures during one's leisure. *Leisure Studies, 32,* 153–171.

Lunney, D., & Ayers, D. (1993). The official status of frogs and reptiles in New South Wales. In D. Lunney & D. Ayers (Eds.), *Herpetology in Australia: A diverse discipline* (pp. 404–408). Mosman: Royal Zoological Society of New South Wales.

Lynch, R., & Veal, A. J. (2006). *Australian leisure.* Frenchs Forest: Pearson Education Australia.

Markwell, K. (2015). *Animals and tourism: Understanding diverse relationships.* Brstol: Channel View Publications.

Markwell, K., & Cushing, N. (2010). *Snake bitten: Eric Worrell and the Australian Reptile Park.* Sydney: UNSW Press.

Morris, R., & Morris, D. (1965). *Men and snakes.* New York: McGraw-Hill.

Natusch, D. J. D., & Lyons, J. A. (2012). Exploited for pets: The harvest and trade of amphibians and reptiles from Indonesian New Guinea. *Biodiversity and Conservation, 21,* 2899–2911.

Ngunjiri, F. W., Hernandez, K.-A. C., & Chang, H. (2010). Living autoethnography: Connecting life and research. *Journal of Research Practice, 6,* 1.

Nissenson, M., & Jonas, S. (1995). *Snake charm.* New York: Harry N. Abrams Limited.

Rainbird, S., & Roswell, J. (2011). 'Literacy nooks': Geosemiotics and domains of literacy in home spaces. *Journal of Early Childhood Literacy, 11,* 214–231.

Roberts, K. (2006). *Leisure in contemporary society* (2nd ed.). Wallingford: CABI.

Ryan, C., & Harvey, K. (2000). Who likes saltwater crocodiles? Analysing socio-demographics of those viewing tourist wildlife attractions based on saltwater crocodiles. *Journal of Sustainable Tourism, 8,* 426–433.

Shine, R. (1991). *Australian snakes: A natural history.* Balgowlah: Reed Books.

Stebbins, R. (1992). *Amateurs, professionals and serious leisure.* London: McGill-Queen's University Press.

Stebbins, R. (2007). *Serious leisure: A perspective for our time.* New Brunswick, NJ:Transaction.

Stutesman, D. (2005). *Snake.* London: REAKTION Books.

Throsby, D., Zwar, J., & Morgan, C. (2017). *Australian book readers: Survey, method and results.* Macquarie Economics Research Papers, 1/2017. Sydney:Department of Economics, Macquarie University.

Tisdell, C., & Wilson, C. (2001). Wildlife-based tourism and increased support for nature conservation financially and otherwise: Evidence from sea turtle ecotourism at Mon Repos. *Tourism Economics, 7,* 233–249.

Tomlinson, A. (1993). Culture of commitment in leisure: Notes toward the understanding of a serious legacy. *World Leisure and Recreation, 35,* 6–9.

Urbanik, J. (2012). *Placing animals: An introduction to the geography of human-animal relationships.* Lantham: Rowman & Littlefield Publishers.

Waldau, P. (2013). *Animal studies: An introduction.* Oxford: Oxford University Press.

Worrell, E. (1966). The unpopular ones. In A. J. Marshall (Ed.), *The great extermination* (pp. 11–90). London: William Heinemann.

Zabriskie, R., & McCormick, B. (2001). The influences of family leisure patterns on perceptions of family functioning. *Family Relations, 50,* 281–289.

Dance with a fish? Sensory human-nonhuman encounters in the waterscape of match fishing

Vesa Markuksela ⓘ and Anu Valtonen ⓘ

ABSTRACT

This study sets out to explore human–nonhuman encounters in the leisure activity of match fishing. Informed by practice theory, studies on the body and the senses, and the human–animal literature, it focuses on analysing the practice-specific, embodied and sensory doings and sayings of both humans and nonhumans during match fishing. The findings from three-year sensory ethnographic fieldwork conducted in Finnish Lapland suggest that human–nonhuman encounters can be characterised as partner dancing. That is, this phenomenon is tantamount to a dance between a fish and an angler taking place in a dancehall of water, in which the weather acts as an orchestra framing the rhythm and tempo of the dance. Considering both fish and anglers, the study emphasises the agential and embodied quality of human–nonhuman encounters. It challenges the dominant position of the human, suggesting a move from anthropomorphism to zoomorphism – animalising the angler in a dance with a fish. The study also provides novel insights into the dynamic nature of a waterscape, highlighting its dual nature consisting of the underwater world and the above-water world. In summary, this study offers a detailed account of the dynamic interactions between humans, nonhumans and the natural environment.

Introduction

Recently, there has been increasing interest in including the nonhuman world within the study of leisure experiences and landscapes (Hughes, 2017). Instead of focusing merely on humans, scholars have begun to explore leisure as a complex, multispecies phenomenon. For instance, researchers have explored human–animal encounters with horses (Dashper, 2017) or dogs (Carr, 2010) and how space shapes these encounters (Cloke & Perkins, 2005). These studies – echoing the wider 'animal' and 'non-human turn' in the social sciences (Despret, 2004; Haraway, 2008) – have been valuable in challenging the prevalent anthropocentric thinking in leisure studies and paving the way toward more balanced between humans, non-humans and environments.

To add to this literature, this study seeks to examine human and non-human encounters in the hobby of match fishing. *Match fishing* refers to the competitive pursuit of angling to catch the heaviest and/or largest fish within a defined time period, according to a specified set of rules (Cowx, 2002). This leisure provides us a fruitful context for further theorising about the nature of encounters between humans, non-humans and the natural environment. It allows us to highlight the sensory and bodily nature of the encounter for both the fish and the human, and to consider the particularities of a waterscape as a context in which and with which these encounters take place.

To study these encounters, we combine practice theoretical stances (Reckwitz, 2002; Schatzki, 2002) with anthropology-zoo-genetic ones (Despret, 2004), also drawing on the literature regarding bodies, senses and movements (e.g. Howes, 2005; Hui, 2012). This study is based on three-year fieldwork conducted in Finnish Lapland, employing a combination of sensory ethnography (Valtonen, Markuksela, & Moisander, 2010) and multispecies ethnography (Dashper, 2017; Maurstad, Davis, & Cowles, 2013).

Consequently, this study illuminates the complex choreography of match fishing that we characterise as partner dancing. Match fishing is a dance between a fish and an angler that takes place in a dancehall of waterbodies. The weather acts as a dance band, orchestrating the rhythm and the tempo of the dance. Our study provides a novel understanding of the dynamic and changing nature of human–nonhuman encounters, highlighting the sensoriality of these encounters, including the non-human sensuousness of the fish.

This paper begins by reviewing the previous literature on human–animal encounters, after which it discusses the practice-theoretical perspective of the study. The methodological section describes the research context and the fieldwork. The analysis, informed by a narrative research approach, illustrates how the encounters taking place in a fishing practice can be described as dancing with a fish. To conclude, we present the contributions of this study to the existing leisure literature.

Previous literature

Social scientists have explored different encounters between humans and water systems as well as various types of water animals and organisms. While Probyn (2016), for instance, has studied aspects of 'human–fish entanglements' in the sea, we explore these entanglements in the context of waterbodies, freshwater fish and sport fishing.

Freshwater angling has gained relatively little attention in leisure research (see, however, Bear & Eden, 2011). Therefore, we lean on studies of hunting because hunting and fishing both are activities that entail a close sensual, embodied relationship between nature and humans (Franklin, 2001; Lovelock, 2008; Mordue, 2009). In these activities, it is not only the co-agency of humans and animals that matters but also the delicate tension that these agencies produce (Franklin, 2001). The heart of hunting is, indeed, a contest based on two sets of senses, that of the human and that of the animal (Marvin, 2005). To date, the focus has been on human senses. Studies demonstrate how hunters seek to sharpen their understanding of animals' senses, as well as their own ability to counter and overcome these animals' senses (ibid.).

In fishing, there is no honing of human senses, because humans do not possess any sense organs that fish do not possess. Quite to the contrary, the fish has sensory organs that humans do not have because the senses of the former are adapted to the water; humans have difficulty seeing and hearing underwater. However, fishermen attempt to alter their own embodiment and senses to match those of the fish. As Mueller's book *Being Salmon, Being Human* (2017) demonstrates, our sense of who we are as humans is mirrored in our lived relationships with other creatures. Mueller (ibid.) sees himself as an embodied mind pondering the lives of other species with very different embodied minds. According to Mueller, his body becomes an arena of confrontation with otherness in a dialogue between the researcher and the salmon. In the same way, Haraway (2008) considers the relationality between human and non-human species, pondering how non-human worlds can change us as we change them.

The vital role of the body and the senses has also been acknowledged in studies of leisure experiences and landscapes (Hughes, 2017). Allen-Collinson and Leledaki (2015), for instance, point out that just 'being-in-the-out-door-world' is a sensuous and intensely embodied act. Humberstone's (2011) study of windsurfing provides another case in point detailing how the body interacts with its natural surroundings via the senses. A recent study of Hughes (2017), for its part, investigates the relationship between the senses and the natural environment by showing how paddling, as a sensuous leisure activity, is deeply haptic and acoustic. Then again, Brown (2017) touches on the leisure environment by investigating how the identity of an offshore sailor is contingent upon being attuned to one's environment via the senses. Studies focusing on sporting activities with animals have, in turn, paid attention to the human-built (obstacle course) or partly natural land-based landscape (eventing course) in which leisure activities such as agility or riding take place (Davies, Maurstad, & Cowles, 2013; Wlodarczyk, 2017). Our activity, match fishing, occurs in a non-built environment, water-scapes, and we address this encounter from a practice-theoretical perspective, as explained in the next section.

Theoretical perspective

Our theoretical perspective draws on practice theories, literature on movement and mobility, and literature that discusses the body and the senses. The form of practice theory on which we lean takes a cultural stance towards understanding social action and social order (Reckwitz, 2002; Schatzki, 2002). With this anti-individualistic approach, the analytic attention is directed to practices that organise and shape individual action. Practices, per se, are conceived as skilful performances from a meaningful repertoire of bodily doings and sayings as well as their accompanying sensations (Schatzki, 2002), including elements of interconnected forms of mental activities, 'things' and their use, background knowledge in the form of understanding, know-how, states of emotion and motivational knowledge (Reckwitz, 2002).

Put differently, practices are assemblages of these elements, and they are 'sets of hierarchically organised doings and sayings, tasks and projects' (Schatzki, 2002, p. 59–63). Each time people synthesise and organise these elements together, they take part in the act of 'practice-as-performance' (Hui, 2012), in which they reproduce the routinised activities of a practice. We treat match fishing as a practice of this kind, as a type of performed, integrative practice (Schatzki, 2002).

Practice-theoretical accounts highlight the embodied nature of practicing, conceiving the body as a carrier of practice (e.g. Reckwitz, 2002; Schatzki, 2002). Commonly, bodily activities are discussed in terms of competences, knowledge and skills. For instance, a bodily activity, such as landing a fish, is something fishermen do as part of their practice with their skilful bodies. Practices – and their attendant rules, values and skills – become thus embodied once they are learnt. In this sense, bodies and practices constitute each other in the embodiment of practices. The body learns to act in a practice-specific way, and in doing so, maintains and reproduces the practice.

The body necessarily is a sensing actor. When enacting bodily activities, the practitioner perceives the world and judges it through the senses. Importantly, the embodied practitioner develops sensory skills that help him or her accomplish the activities at hand (Vannini, 2011). The senses are not only a means of apprehending physical phenomena, but also are invested with cultural values and meaning (e.g. Classen, 1997; Howes, 2005).

Embodied practitioners perform acts of movement; even more seemingly static acts, or even motionless ones, entail movement of some sort (Hui, 2012). Mobile activities are particularly central to our encounters with nature, as the body and its particular 'equipment' must anticipate and react to the continually changing environment (Humberstone, 2011). Through pondering the micro-movements of the body, we can analyse the dance, capturing its sequences, rhythms and routines. The movements of the body and its limbs allow the sensory inspection and observation of our surroundings.

Our treatise acknowledges that non-humans also are carriers of practice. In troll fishing, the non-human world actively contributes to the perpetuation of practices. Fish in a body of water swim their routes below the surface and propel their sensing bodies through the water. Fish also have their own bodily routines and rhythmicity (e.g. eating, sleeping, mating) that form under-water patterns, constituting the secrets of the world of the fish. Likewise, the weather and water have the capacity to enable, constrain, direct and redirect human practices and thereby shape forms of coexistence and action (Rantala, Valtonen, & Markuksela, 2011). In this sense, non-humans – fish, weather and water – are treated as active elements within practices that can be routine or, conversely, can involve random and unusual, emergent happenings.

The Pickerian metaphor of a 'dance of agency' is apt for our analysis: 'we act in the world, the world acts on us, to and from, in a dynamic process' (Pickering, 2017, p. 4). All the agents (anglers, fish and waterbody) are unpredictably and emergently transformed, and the agency performed during the dance is distributed among those who take part in fishing practices (Mattila, Mesiranta, Närvänen, Koskinen, & Sutinen, 2018). There is, however, an imbalance of power between practitioners. In our case, the fish, on some occasions, have little possibility of refusing to take part in that practice.

Research methodology

To empirically investigate the anglers' embodied and sensory encounters with the fish and the water, this study leans on sensory ethnography (e.g. Classen, 1997; Howes, 2005; Valtonen et al., 2010). This enables us to explore the senses in action in the immediate settings within which the activity takes place. In line with the practice–theoretical lens, the analytic focus is directed toward the ways in which the senses play a part in the performance of practices and in subsequent encounters with the social and natural world.

The lead author of this work conducted three-year multi-site ethnographic fieldwork in Finnish Lapland by participating in 22 match fishing competitions, which lasted from eight to 24 hours and took place in a variety of freshwaters. Each year, the trips lasted from June to November. Therefore, a wide spectrum of Lappish weather – from heat to sleet – was experienced. The scrutinised mode of match fishing is trolling, a method of fishing in which some form of bait is drawn on a line through the water from watercrafts by two to four anglers. Competitive trolling provides an arena for the investigation of human and nonhuman encounters and grazing. The angler directly and indirectly (thorough technology) encounters many species of fish and other non-human actors, such as birds and flora (e.g. the unwanted touch of bottom plants or snags), not to mention mosquitos. The author belonged to a local trolling club whose members – all men – participated in fishing competitions. The very idea of a fishing competition entails several stereotypical masculine values such as domination and conquest over nature (Adkins, 2010; Birke, 2012). Indeed, every fishing competition exudes the human desire for challenge and racing. It is performed on a 'showdown' stage, where anglers gather in a competitive manner to perform 'dance choreographies' that entail skilful intra-actings of all the entangled elements of a practice. The competition also represents a moment when one is not supposed to fail; just 'asking to dance' is not good enough – anglers 'gotta dance'.

During the competitions, the lead author was one angler among others, either in his own boat or as a crewmember in other anglers' watercrafts. The researcher thus immersed his body in fishing activities and, for that matter, moments of paused mobility (Sheller, 2014). This allowed him to make participant observations 'from the inside'. He also carefully 'listened' to his own body, reflecting this bodily knowledge within the fishing practice and comparing it with the ways that a fish senses. The latter turned out to be the most challenging part of the fieldwork: the researcher had to be simultaneously aware of his own human sensory embodi-ment and put effort into detaching himself from human sensations to grasp the sensing body of the fish.

The data were generated via participant and non-participant observations (written head notes, field notes and diaries), informal discussions on the water and onshore with other anglers, visual materials (e.g. photos, video clips) and autobiographical stories. Most commonly, the data were gathered via observations and technical apparatuses. For instance, closely following the screen of a fish-detecting sonar system or depth camera allowed the ethnographer to follow the movements of the fish. While doing so, he was constantly asking himself questions. What did the fish just do? Why did the fish do it like that? How did the fish do it? What senses were involved in these actions? Gradually, the researcher was able to create a sensorial connection and understanding between the studied fish and the human, as well as the associated micro-mobilities of the body, bodily rhythms, and motions (Sheller, 2014). Thus, interconnection can occur without actual 'contact zones' via technologised observation. This 'distant' way of familiarising differs from Haraway's (2008, p. 3) view of 'learn to be worldly from grappling with', in which familiarisation happens in close contact with humans and non-humans. Apparently, distant ways may reduce the 'unknown' characteristics of fish, and the non-human other can become close and familiar (Bear & Eden, 2011).

The analysis took the form of a practice-theory-informed, data-driven narrative analysis (Gubrium & Holstein, 2008). During and after the fieldwork, the author wrote autobiographical narratives that combined his experiences and the observations of human and non-human others, discussions and reflections. They form the 'LEGO bricks' of a grand narrative. The results of the analysis are also represented in the form of a narrative – short vignettes, to be more precise. These vignettes are chosen to illuminate the various aspects of dancing with a fish, as the following section shows. They are written 'from the inside' of the cultural practice of fishing that is competitive and male-centred by its very nature.

Human–nonhuman encounters in the waterscape of match fishing

The wind has calmed down, and the heavy rain has stopped. The air is moist, and the sky is full of shades of grey. A gentle wave ripples the surface of the lake. The wind has blurred the watercolour. Bouncing gently, our watercraft moves at a slow trolling speed above the water. I stand up from the bench and glance around. There are no isles or shoals in sight. I can't spot any fellow anglers, either, only flat water as far as the eye can see. One could say that we are in the middle of nowhere, but this is not the case. We are exactly where we want to be. We are in our fishing grounds – our secret water fishing spot.

I glance down at the sonar's large colour screen, which displays a diverse, lively underwater landscape. The sonar outlines a basin wall, which sinks steeply into an abyss. The sonar adduces the changing features of the bottom of the lake. The different colours on the screen also indicate the depth of the water and the various water temperature layers, or thermoclines. It is this basin wall and these changes in depth and water temperature that are tempting to life, particularly fish. The sonar outlines many fish underneath us. There are large, moving schools of small fish, and nearby, there are swarms of larger fish, presumably pikes, the catch we seek.

The fish are mobile, but they stay in a relatively small area. Nonetheless, it takes some time to locate promising stock. Since this discovery, we have been crisscrossing this distinct area, waltzing various lures near the fish, wooing and politely 'asking' the fish to take a bite.

According to the sonar, the water temperature is +14°C. The pike love this kind water. It's the pikes' hunting season, so why don't they strike? I'm stupefied. Is it the changed weather? Helplessly, my mind goes back to yesterday. At this time and in the same spot, I was holding a rod, and on the end of the line was a large pike. After the hooking, we had a long fight. The pike had run and dived. I had pumped, released and provided pressure, time after time. I was almost closing the deal, ending the dance, when the pike suddenly jumped and managed to release itself from the hook. Its large tail waved goodbye as it dove into the depths.

Today, I have the same lure in the water. Come on, what's the matter? You should know this fish by now! Feel it! Sense like a pike!

This vignette describes one scene in a day of competitive fishing, illustrating how the fishing practice is comprised of encounters between the fish and the anglers within the waterscape. It also illustrates the ways in which the movements of the human body and those of the fish, as well as both their senses, are central elements in the accomplishment of an angling practice. It highlights that the key dynamics of the mobility and movements of anglers are based on the senses. The angler seeks to understand the movements and senses of the fish by sensing the weather and the changing waterbody, either directly through his or her body or via technological devices such as sonar.

As this vignette indicates, the different encounters that make up fishing practice can be characterised as a kind of dance between a fish and an angler. This dance has elements of open position partner dancing, ballet and even line dancing. Our analysis brings forth three components of dance in competitive match fishing: (1) fish and angler as dancing partners, (2) the waterbody as a dancehall and (3) the weather as orchestral accompaniment.

Dancing partners

When fishing, the angler does not necessarily need to know his or her dance partner beforehand. However, to 'hook' with a partner in this dance, he or she must first be properly introduced to the prospective dancing partner. This introduction stipulates that he or she can locate the fish. In this quest, watercrafts are gliding here and there, performing crisscrossing choreography above the water, and looking constantly below the surface into the underwater world. Anglers seek to dive into the abyss metaphorically because the aquatic is a part of the practical understanding of fishing practice. The aquatic world is also the 'neighbourhood' of the fish (Figure 1).

To find a dance partner, the competent angler attempts to familiarise himself or herself with the fish. The fish has a lifecycle of its own, as well as seasons of activity and diurnal rhythms. They eat and sleep. The times and places of these activities vary between species and between individual fish. A competent angler may specialise in one specific species (e.g. pike), but in match fishing, it makes sense to become acquainted with all 'catch' species in the competition at hand and, more broadly, with the ecosystem of the fish. Unlike the vignette above, anglers seldom have the privilege of spending time exclusively on their own in a 'spot'. Usually, there are fellow anglers offering lures to fish; indeed, one crew can have up to a dozen lures in the water. In a way, trolling crews make up their own swarms, next to existing fish schools.

When the trolling watercrafts with their fascinating lures arrive in the 'neighbourhood' of the fish, a disturbance is caused. In the beginning, before the dance and partnering, there may be some reluctance in the water. The fish may move away from the angler as he or she attempts to move toward the fish. When approaching the fish, the angler endeavours to sense like a fish and to anticipate its location, movements and intent. The fish, like many other animals, can feel curiosity

Figure 1. Crisscrossing above Lake Kilpisjärvi.

(Burghardt, 2005). Using different lures with various sensorial stimuli, anglers strive to rouse the curiosity of the fish. It is curiosity and sensorial cues that drive both humans and fish down new paths, in this case to the incipient partner dance.

The senses of a fish diverge from those of a human. The fish does not have external ears, but anglers know that the fish hears with the inner ear (Franklin, 2001). Making the right kind of noise can be the key to getting a fish to strike. Thus, the angler often makes a lure audible by attaching various sound effects, and he or she also knows that the noise from banging on the bottom of the boat will conduct extremely well through the water.

Anglers commonly try to learn about the senses of the fish, for instance, through magazines and meetup e-groups. For example, the fish uses the sense of smell to search for nutrition, detect enemies, and identify its own species, and it uses taste to identify the quality of nutrition (Franosch, Hagedorn, Goulet, Engelmann, & Van Hemmen, 2009). While the fishing industry has developed pheromonal gels that can be attached to a lure to attract fish, other more traditional means of enhancing lures are also used, such as dipping them in a fish broth or spitting human saliva onto the lure.

Finland's freshwaters are not particularly clear. Therefore, the area that fish can see is limited – the weather and the light affect the ability to see (Bleckmann & Zelick, 2009). Colours disappear as one goes deeper into the water column. Thus, the fish can spot the colours of the lure only fairly nearby in deeper waters. Anglers also tend to use visible colours in their lures, especially in muddy waters, on cloudy days or in deep waters. They help the fish to see better.

The fish also has a unique sense, known as the *lateral line*. With the aid of this sense, the fish can move in dark waters and at times when it is difficult to see (e.g. night or winter). It is a type of sonar that fish use to orient themselves, notice obstacles and hunt (Bleckmann & Zelick, 2009).

A competent angler wants to have at least one distinctive lure that will stand out from the dozens of others being offered. Anglers seek to appeal to all the senses of the fish. Hence, they create sensual stimulus bundles by bringing their lures to the right depth, close enough to the fish. In this way, the fish can see the colours; hear and feel the vibrations of the lure, which imitates the movements of a wounded quarry; and smell alluring scents.

The offered lure can be seen as a petition, as a way of asking a fish to dance. When the fish bites the lure, dancing partners are hooked, and the dance can finally begin. It is a dance of belonging that may turn out to be the last dance of all. This is partner dancing in an open position, in which the partners are connected without bodily contact. The dancing partner, the fish, is 'at arm's reach'. The angler has a 'grip' on the partner through the rod and line. This phase of the dance emphasises the sense of touch and the dance's kinaesthetic dimensions. In fact, the partners usually do not see one another, due to the colour of the water, the distance between them or the depth. The existence of the dance is expressed through bodily movements, and these are sensed by both parties through the fishing line. On the angler's side, the sense of hearing is 'knotted' to the kinaesthetic movement. The angler can hear the loud sound of the reel's fishing drag. Simultaneously, he or she can feel how hard the fish pulls on the line. This gives the angler a hint of the size of the fish. The other partner, the fish, receives sensory perceptions via the line pressure using the sense of touch and the lateral line.

This partner dance is not strictly choreographed. At least, the other dance partner has not 'read the directions'; even if the angler has written such. There are no fixed leader and follower roles. Usually in dancing, the leader suggests by one's lead the figures that will be executed. The leader has the privilege of maintaining the rhythm and deciding what figures one will lead the other dancer into. The angler, on the one hand, strives to lead by maintaining a tight line, reeling and pulling with the rod in a smooth manner. On the other hand, when a fish is making a run and is in charge, the angler extends his or her line and allows the fish to lead and perform its own dance manoeuvres.

In this form of dance, partners lead and follow one another, switching back and forth. They compete for the leading role. For instance, the fish might wish to be the principal dancer and to perform adagio, leaping solos to stand out in the *pas de deux*, the dance duet. The angler is often in the role of the follower, who does not have any idea what the leader, the fish, will do next.

There is always the possibility of failing in moments of resistance and evasion during this dance. The angler's knowledge of the special conduct of a given fish species reduces the risk of stamping on his or her dance partner's toes.

'Fighting' and 'landing the fish' are the closing phases of this partner dance. Then, anglers must adjust their movements, considering those of their fellow fishermen who are sharing the water-craft's relatively small inner spaces. In dance terms, this could be 'a rolling grapevine' move or a form of line dancing. Along with performing such dance moves, anglers must also manoeuvre the watercraft to enable practice performance.

This partner dance can end in many ways. The fish can get away. Its runs and leaps, as dance manoeuvres, can be successful due to line breaks or detachment from the hook. Then, the fish is the 'winner of the dance competition'. Alternatively, the last dance can occur above the water, at the bottom of the vessel. This is a dance of death, and the inanimate fish will never dance again. The following excerpt from a field diary illustrates one such scene.

> I look at the bottom of my watercraft. Yes, the big, dead corpse of a pike lays there. It's easy to tell: it's my record, the biggest fish I've ever caught! I hear my dad's rejoicing voice, 'Yes, yes, YES'! Still, at this peak moment, I am mostly puzzled. My mind somehow can't join my father's exultations. I ask myself, 'Where is the joy and elation? Why don't I feel satisfied? Why do I feel so numb'? I take one more look at the big fish, and remorse starts to grip me.

After the tumult of the fish dance, it is not unusual for the angler to feel remorse for the fish and its destiny. Perhaps the angler is hoping that there could be an alternative end, a happy one, a win–win situation in which both dance partners could depart as living and more experienced 'dancers'.

In a sense, the deceased fish can still be in touch with the living world through its dead body. For instance, in trophy shots, the angler holds the fish's corpse, which is memorialised in pictures. There is yet another bodily resurrection, or magical encore, namely, the weighing of the catch. The deceased fish can lead the angler to glory on the podium. This can also be seen as reinforcing both the supremacy of the human and the oppression of the animal (Birke, 2012). Finally, on certain occasions4 anglers make food from their catch and gather for shared meals – an act that honours both the fish and the anglers. Furthermore, eating the catch represents another form of bodily entanglement between the angler and the fish (Probyn, 2016).

Dancehall

The above dance encounter takes place in a certain waterbody, a dancehall. In this study, the dancehall is a river or lake whose surface area and depth topography change repeatedly. These waters may have reefs or islets, and their landscape is volatile. No inner or outer homogeneity is to be found. The dancehall's general features and topographical structure guide life under the water and the praxis of the fish, also determining the possibility of the fish surviving in the water. The specific features, such as the basin wall, create an essential framework for a functional micro-ecosystem and the well-being of the fish.

As illustrated in the vignette, the basin wall is an anomaly in the otherwise monotonic structure of the bed of the water, gathering a magnitude of organisms and providing the basis of a rich food chain. For instance, larger predatory fish seek schools of small fish that, in turn, find their nutrition in the basin area. Situational changes in the surroundings of the waterbody can also threaten the lives of the organisms in it. For example, long-lasting hot weather and a high-pressure area can warm the water too greatly. If thermoclines disappear and there is only warm water left, the organisms left in it are threatened, as their oxygen may run out.

Waterbodies' common and situational circumstances define whether the dance will flow. Is the dancehall's floor slippery or frictional? Is there space, or is it crowded? Before the dance 'match up', trolling watercraft wander over every part of the dancehall, in the corners and in the midst of the floor, looking for a dance partner. In this search, anglers' own bodies are moving in the vessel, and the same time, their bodies are moved by the watercraft.

The watercraft is moving simultaneously in the landscapes over and under the water. The non-human dancehall's unique feature is this vertical depth dimension. There is an aquatic, under-water landscape that is impacted by the conditions above it. For instance, the air pressure and temperature, wind and weather (e.g. sunny or rainy) determine the state of the water. Are there surface waves or high-wind waves? What kinds of thermoclines are present?

The dancehall can be crowded for purely topographical reasons. Moreover, the number of trolling watercrafts can cause jams. Dance cavaliers may come too close to one another. The situational elements of the waterbody, such as whether there is calm or surf, may operate as friction in the dancehall. Therefore, trolling crafts' motions above the water may not be smooth. The vessel may be bouncing and tossing around from one wave to another. Metaphorically, the watercraft represents dancing shoes. There may or may not be friction between the shoes and the dance floor during the glide above the water. The dancing shoes can also be slippery and cause the dancer to fail. Next, consider an excerpt describing the act of landing a fish:

> The fish had made several runs, leaped and dived. Now, I can feel less drag, less tension in the line. The fish stops running, and I manage to reel in the slack line. The fish finally surfaces and flashes its flank. Softly, I mutter, 'Reckon it's about time'. My dad stands up from the cockpit on the left; he knows what I mean. Dad is the net-man. It's his job to land the fish. Perhaps due to the long wait, my dad rushes past to retrieve the net from the right side of the boat. He nudges me as he goes by. Dad is a tall man, a big fish in a small pond. I almost drop the rod and fall overboard. Without saying a word, my dad returns with the net. I feel yet another impact. This time, it's the landing net's hoop and the rubber mesh that grazes my face. Meanwhile, the waves toss the boat unexpectedly. There is no one to hold the wheel, no one to warn us about the coming waves. I lose my balance and start to fall...

This excerpt shows how both anglers' movements in the craft and the water's sudden practices (the coming waves), which move the watercraft and the bodies within, can cause friction and stumbling, that is, stepping on toes.

An angler encounters the water and its changing nature via the formal Western senses: sight, hearing, smell, taste and touch and its kinaesthetic constituents. In these sensory encounters, the embodied doings and collectively shared meanings of the space are produced and attached to it. In this way, a neutral 'space' is shaped into a specific dynamic 'place', a multifaceted sensory waterscape and, perhaps, into a 'holy ground' of matchmaking.

Dance orchestra

Dances are distinct from each other in the same way that rhythms of different types of music are. The rumba is danced differently from the samba or swing. In the 'dancelike' practice of fishing, the weather acts as a dance orchestra, providing music for the dancers. The rhythm of the music, in turn, orchestrates the circumstances of the dancehall, that is, the waterbody. It moulds the context in which both fish and anglers saunter. The rhythm that the weather provides organises the match fishing praxis. Is it possible to fish? Where should the watercraft be piloted? The weather above the waterline also shapes the weather below the waterline. The 'weather below' (e.g. the water pressure and temperature) affects the behaviours of predatory fish and fish schools, including their whereabouts and movements. For example, in times of high water pressure, fish gravitate toward the colder layers of water on the bottom. There, they are inactive, and feeding activity is low. In the angler's practical understanding, such fish are said to have 'lockjaw'.

The weather provides the basic beat for the praxis of fishing, and it also affects the dancing style of the potential partner dance. The weather and its various material forms, especially the wind, stipulate the rhythm of the dance. Is it a hot-tempered and hectic rumba of high tides or a cheek-to-cheek dance at a slow tempo in calm waters?

Figure 2. Rumbling rumba of high waves.

Rhythm is 'the perception of an order' (Fraisse, 1982, p. 151). This definition implies that we can predict or anticipate what will come next in a sequence. In arrhythmia, in contrast, such an order is not assessed directly (ibid.). In match fishing, this arrhythmia emerges when anglers are water bound and something unexpected happens in terms of the weather. Competent anglers will attempt to avoid this and seek to sense the upcoming weather and water circumstances. That is, they want to be weather-wise (see Rantala et al., 2011).

For anglers, the preparation for the dance and journey to the dancehall begin with a consideration of the prospective weather. Anglers do not rely solely on meteorological radar forecasts and other weather prediction techniques. They also utilise prediction by observation, with which they try to interpret the 'sensorial cues' of nature and attach them to experiences and knowledge, both their own and those of others. This implies, for instance, interpreting the colours of the water, waves and clouds, as well as air odours. Observations of certain plants, such as trees (the movement of sprigs and the swaying of the trees) or certain animals' appearance and behaviours (e.g. swallows' flight altitude, with a low altitude indicating rain) can also be a basis for predicting the weather.

The prevailing weather conditions constitute a central feature of the waterscape and play a crucial role in shaping the sensory elements of the scape in question. Next, consider an excerpt from a narrative describing fishing in challenging changing summer weather conditions:

> Our watercraft is moving quickly to the competition centre. We are fleeing from the upcoming storm. Longish waves begin to form due to the wind. Our aluminium craft bounces from one wave to another. I have to hang onto the railing not to fall overboard. Water is pounding against the watercraft, splashing inside, making us wet and even colder. The wind is getting more intense. I feel it blowing more forcefully against my face, and at the same time, the wind whistles in my ears. The waves instantly become even longer. With another bounce, my head is flipped backwards. Suddenly, I am gazing at the sky, where more and more dark clouds are gathering. It smells like rain. The wind begins to wail, and instantly, the rain streams down.

This excerpt illustrates that a sudden weather change may surprise even relatively skilled anglers who have tried to anticipate the forthcoming weather. Coping with sudden arrhythmia typically requires altered fishing practices, for example, seeking the shoreline or some other shelter from the waves. Sudden arrhythmia also affects assorted senses. Usually, proprioceptive (sense of balance) sensations are emphasised. Besides sudden arrhythmias, there are also seasonal changes in the weather. In our context, summer in Finnish Lapland, the weather can range from heat to sleet. The weather's rhythm also determines the angler's dress code in the dancehall. On rainy days, a floating suit is needed. In the sun, the dress code is more casual, and a swimsuit is allowed.

Discussion and conclusion

This study has provided an empirically grounded analysis of the ways in which humans and nonhumans, anglers and fish, interact bodily and sensorially in the waterscape in the midst of a match fishing practice. It was suggested that this multifaceted encounter is best described as a form of partner dance, as a dance with a fish. This metaphor enables us to highlight the embodied, dynamic and changing nature of the encounter and focus on the agential qualities of non-human entities. Fish and waterbodies are not passive targets or platforms for actions but active constituents of the action. Our study thereby contributes to the existing leisure studies literature, which has begun to include the nonhuman world, in the following ways.

First, our study offers novel insights into the ways in which the bodies and senses of both the animal and the human play a role in the encounter. The first part of the analysis detailed how the anglers search for a dance partner in the dancehall. In this search, the angler is attempting to adopt, bodily and mentally, the same wavelength as the fish. This can be understood as an Ortega-y-Gassian-process, which refers to 'being open to the animal'. For instance, anglers attempt to adopt many of the practices of fish, such as eating, and set their human practices aside. Our analysis further elaborated on the actual dance and the ways in which partners are in touch through the material elements of line and rod. While perhaps ephemeral, the encounter carries meanings of deep connectedness. The non-human and the human exchange intense and extremely sensual moments. The angler can sense the fish, as an individual, and connect this sensation to the fish's species-specific attributes. In sensing like a fish, the angler is, in a way, becoming an animal, not in terms of understanding the fish's inner experiences (Mueller, 2017), but in terms of the fish's practical understanding. This may not be merely a one-sided change, because the fish may also be pulled into becoming a man (Wlodarczyk, 2017). Then, the fish has the prospect of becoming aware of the angler's intentions and practices, of learning how a man acts. Thus, the fish is acting not on but with human anglers (Haraway, 2008).

In the dance, the fish is considered a discerning artist of the waterworld. For example, it can choose whether to strike the lures offered. Thus, the fish has pivotal agency. This somewhat deviates from the existing literature on human–animal relationships and its latent humanocentrism. For instance, Dashper (2017) argues that, while horses in human–horse relationships can demonstrate some agency by choosing to interact with or ignore humans, their choices are 'bound by the human-centric context in which these interactions take place' (ibid., 4). In our case, there are various levels of freedom. Our findings also differ from the previous literature on hunting (e.g. Marvin, 2005), which concerns the honing of a (hu)man's abilities to defeat an animal.

Second, our analysis of the waterscape contributes to the existing knowledge of leisure landscapes. In her study of surfing, Humberstone (2011) emphasises that the landscape structure in water-based activities is important, because the 'seascape' is always mobile and fluid. Humberstone provides an account of how this changing scape evokes human emotions – it is about feeling-in-the-place. Our analysis, in turn, demonstrates how the ever-changing waterscape is coped with via adjusted and embodied sensory practices – for us, it is doing-in-the-place. In line with our approach, Humberstone (2011) illuminates the animate and sensory characteristics of the waterscape. However, she is concerned only with the world above the water. We suggest that the waterscape structure is divided into interconnected above-water and underwater 'landscapes'. By highlighting this dual nature of the waterscape, our analysis offers a novel perspective, compared to studies of landscapes framed by the notion of gaze and focusing commonly on the aboveground landscape (Urry & Larsen, 2011). We also continue the discussion concerning the active agency of natural elements over leisure places (Cloke & Perkins, 2005) by pointing out the dynamic relationship between the above-water and underwater worlds. The weather above the waterbody affects the conditions below, where the fish are. Then again, the fish underwater guide the actions of anglers in the above-water landscape.

Third, our understanding of weather's role as a force of nature adds to pending discussions of non-human agency in leisure activities. In their work, Rantala and her colleagues (2011) articulate how weather guides the material and corporeal human practices that take place in natural environments. Our analysis continues this insight by discerning how weather also governs the practices of non-humans. For instance, changing weather above the surface, such as a torrent or storm, causes arrhythmia in the waterbody (e.g. tumultuousness). This, in turn, breaks up the practice routine of fish and anglers. We do not treat sudden weather changes as a mere disruption (see Cloke & Perkins, 2005). For us, arrhythmia is a new situation calling for situated weather practices. The arrhythmia evoked by the weather and its agency in leisure activities and the waterscape contributes to existing discussions of the relationship between humans and non-humans. In fact, we suggest that a waterbody's arrhythmia can be treated as a special kind of landscape.

Overall, we have used the metaphor of dance to describe the relational entanglement between fish, angler and waterbody in a fishing competition. The metaphors are not without problems, as they simultaneously highlight and hide aspects of the phenomenon (Lakoff & Johnson, 2003). The metaphor of dance highlights the very logic of the fishing practice, as it clarifies that fishing is a pair activity: it takes two to tango.

Furthermore, the given metaphor enables one to highlight the thoroughly embodied nature of the practice. Both dancing and fishing are bodily activities in which one must account for the bodily movements of others, in this case, the bodily movements of the non-human fish. As dancers, anglers must improvise, often because of suddenly changing weather conditions.

While the metaphor of dance arguably is human-centric, it still accords the fish a more 'equal' position in that the fish is understood as a partner that has the power to guide and frame the angler's actions. Dance – especially a duet – is a gendered activity. The practice of fishing is predominantly a male-centred activity and involves logic as well as sayings and doings that are commonly interpreted as masculine, such as competing, fighting and winning. Therefore, the metaphor of dance is hardly appropriate for an analysis that seeks to uncover wider sociopolitical implications of the thoroughly gendered nature of the more-than-human entanglements formed around fishing, as cogently discussed by Probyn (2016). Yet, it captures adequately the cluttered encounter of the human–animal relationship in the practice of match fishing, during which both parties are bodily knotted to each other in a swaying motion. As one angler explained amidst the whirl of fighting: 'One step forward, two steps back. Heck, this fish is leading this waltz. I'll bet this will go to the last call'.

Disclosure statement

No potential conflict of interest was reported by the authors.

Funding

The authors have not received no funds for the completion of this work.

ORCID

Vesa Markuksela (iD) http://orcid.org/0000-0002-1106-4830
Anu Valtonen (iD) http://orcid.org/0000-0001-8940-1254

References

Adkins, T. (2010). *Fishing for masculinity: Recreational fishermen's performances of gender in the rust belt*, Master of Arts Thesis (Sociology), Kent State University.

Allen-Collinson, J., & Leledaki, A. (2015). Sensing the outdoors: A visual and haptic phenomenology of outdoor exercise embodiment. *Leisure Studies, 34*(4), 457–470.

Bear, C., & Eden, S. (2011). Thinking like a fish? Engaging with nonhuman difference through recreational fishing. Environment and planning D. *Society and Space, 29*, 336–352.

Birke, L. (2012). Animal bodies in the production of scientific knowledge: Modelling medicine. *Body & Society, 18* (3–4), 156–178.

Bleckmann, H., & Zelick, R. (2009). Lateral line system of fish. *Integrative Zoology, 4*, 13–25.

Brown, T. (2017). The offshore sailor: Enskilment and identity. *Leisure Studies, 36*(5), 684–695.

Burghardt, G. M. (2005). *The genesis of animal play: Testing the limits*. Cambridge, MA: The MIT Press.

Carr, N. (2010). Animals in the tourism and leisure experience. *Current Issues in Tourism, 12*(5–6), 409–411.

Classen, C. (1997). Foundations for an anthropology of the senses. *International Social Science Journal, 153*, 401–412.

Cloke, P., & Perkins, H. C. (2005). Cetacean performance and tourism in Kaikoura: New Zealand. Environment and planning D. *Society and Space, 23*, 903–924.

Cowx, I. G. (2002). Recreational fishing. In P. B. J. Hart & J. D. Reynolds (Eds.), *Handbook of fish biology and fisheries* (Vol. II, pp. 367–390). Oxford: Blackwell Science.

Dashper, K. (2017). *Listening to horses: Developing attentive interspecies relationships through sport and leisure*. Society and Animals, *25*(3), 207–224.

Davies, D. L., Maurstad, A., & Cowles, S. (2013). Riding up forested mountain sides, in wide open spaces, and with walls. *Developing an Ecology of Horse–Human Relationships, Humanimalia, 4*(2), 54–83.

Despret, V. (2004). The body we care for: Figures of anthropo-zoo-genesis. *Body & Society, 10*(2–3), 111–134.

Fraisse, P. (1982). *Rhythm and tempo* (pp. 149–180). Orlando, FL: The Psychology of Music Academic Press.

Franklin, A. (2001). Neo-Darwinian leisure, the body and nature: hunting and angling in modernity. *Body and Society, 4*, 57–76.

Franosch, J.-M. P., Hagedorn, H. J. A., Goulet, J., Engelmann, J., & Van Hemmen, J. L. (2009). Wake tracking and the detection of vortex rings by the canal lateral line of fish. *Physical Review Letters, 103*(7), 78–102.

Gubrium, J. F., & Holstein, J. A. (2008). Narrative ethnography. In S. Nagy Hesse-Biber & P. Leavy (Eds.), *Handbook of emergent methods*, 241–264. New York, NY: Guilford Publications

Haraway, D. (2008). *When species meet*. Minneapolis: University of Minnesota Press.

Howes, D. (2005). An introduction: Empire of the senses. In D. Howes (Ed.), *Empire of the senses: The sensual culture reader* (pp. 1–20). Oxford: Berg.

Hughes, C. (2017). A song of the paddle: Haptic aesthetics of canoe travel in the English Lake District. *Leisure Studies, 37*(3), 268–281.

Hui, A. (2012). Things in motion, things in practices: How mobile practice networks facilitate the travel and use of leisure objects. *Journal of Consumer Culture, 12*(2), 195–215.

Humberstone, B. (2011). Embodiment and social and environmental action in nature-based sport: Spiritual spaces. *Leisure Studies, 30*(4), 495–512.

Lakoff, G., & Johnson, M. (2003). *Metaphors we live by*. London: The university of Chicago press.

Lovelock, B. (2008). *Tourism and the consumption of wildlife: Hunting, shooting and sport fishing*. New York: Routledge.

Marvin, G. (2005). Sensing nature: Encountering the world in hunting. *Etnofoor, 18*(1), 15–26.

Mattila, M., Mesiranta, N., Närvänen, E., Koskinen, O., & Sutinen, U. M. (2018). Dances with potential food waste: Organising temporality in food waste reduction. In *Time & society*, July 2018, Online First. doi:10.1177/0961463X18784123

Maurstad, A., Davis, D., & Cowles, S. (2013). Co-being and intra-action in horse-human relationship: A multi-species ethnography of be (com)ing human and be (com)ing a horse. *Social Anthropologie, 21*(3), 322–335.

Mordue, T. (2009). Angling in modernity: A tour through society, .nature and embodied passion. *Current Issues in Tourism, 12*(5–6), 529–552.

Mueller, M. L. (2017). *Being Salmon, being human. Encountering the wild in us and us in the wild.* White River Junction, Vermont: Chelsea Green Publishing.

Pickering, A. (2017). The ontological turn: Taking different worlds seriously. *Social Analysis, 61*(2), 134–150.

Probyn, E. (2016). *Eating the ocean.* Durham: Duke University Press.

Rantala, O., Valtonen, A., & Markuksela, V. (2011). Materializing tourist weather: Ethnography on weather-wise wilderness quiding practices. *Journal of Material Culture, 16*, 285–300.

Reckwitz, A. (2002). Toward a theory of social practice: A development in culturalist theorizing. *European Journal of Social Theory, 5*(2), 243–263.

Schatzki, T. R. (2002). *The site of the social: A philosophical account of the constitution of social life and change.* University Park, PA: The Pennsylvania State University Press.

Sheller, M. (2014). The new mobilities paradigm for a live sociology. *Current Sociology Review, 62*(6), 789–811.

Urry, J., & Larsen, J. (2011). *The tourist gaze 3.0.* London: Sage Publications Ltd.

Valtonen, A., Markuksela, V., & Moisander, J. (2010). Sensory ethnography in consumer research. *International Journal of Consumer Studies, 34*(4), 375–380.

Vannini, P. (2011). *Senses, Encyclopedia of Consumer Culture, by Dale Southerton.* New York and London: Sage Publications, Inc.

Wlodarczyk, J. (2017). Be more dog: The human–Canine relationship in contemporary dog-training methodologies. *Performance Research, 22*(2), 40–47.

Shared spaces on the street: a multispecies ethnography of ex-racing greyhound street collections in South Wales, UK

Kerry L. Sands ⓘ

ABSTRACT

Every year in the towns and cities across Wales, humans and ex-racing greyhounds come together onto the streets in the name of charity. These public fundraising and awareness-raising events offer unique opportunities to unravel the reciprocal strands of canine and human experience which coalesce in leisure contexts. I conducted ethnographic research culminating in an autoethnographical account of four of these events, to determine how these encounters might be constituted for multispecies participants. Humans intra-acting with greyhounds on the street experienced an affective state; what I have termed an 'auric imprint', which prompted further exchanges of an emotional, economic and practical nature. Ideologies about greyhounds were underpinned by cultural understandings, situating these dogs as 'not-like-other-dogs' and resulting in a perpetuation of their representation as objectified, homogenous entities, and limiting their recognition as individual dogs. A practical embodied approach, inspired by discourse supporting attentive awareness to non-human animals, was discussed as a mechanism for enabling the recognition of greyhound subjectivities at street collections. Future research might examine how street collections may provide spaces of insight where human behaviour change for non-human animals can be tested and enacted.

Greyhound racing in Wales

Greyhound racing has a long history in the working-class strongholds of the South Wales Valleys, UK. At one time, over 20 tracks were in operation, but now only one independent or 'flapping' race track remains; the Valleys Greyhound Track in Ystrad Mynach, Caerphilly, or simply 'Valleys' as it is known locally. Greyhound racing is a contemporary leisure pursuit which, in the UK, has two modalities; registered racing (colloquially known as racing 'under rules'), and independent racing (known colloquially as 'flapping'). Registered racing covers those tracks and activities licensed and regulated by the industry body, the Greyhound Board of Great Britain (GBGB). Currently, 22 GBGB licensed tracks operate in the UK; 21 of these are in England with only one registered track in Scotland. Independent racing ('flapping') is also represented in Atkinson & Young's 'greyhound racing figuration' (2005, p. 352, 2008, p. 85) and historically has its roots in the ex-mining working class communities of Wales, Scotland the North East of England (Huggins, 1948). In addition to the 22 GBGB licensed tracks in the UK, five 'flapping' tracks continue to operate, which includes the one remaining track in Wales.

Established in 1976, race nights at 'Valleys' operate twice weekly, assuming enough entries of dogs to make meetings viable. Entry to the venue is £5 for adults; children go free. Trainers/owners also have to pay to enter although they may enter dogs into the races for free. Prize money is £20 for the winner. During the track's heyday of the 1970s and 80s, the number of bookmakers at the track were in double figures, whereas today, only two remain and bets can be placed as low as £1. The venue is marketed via a dedicated Facebook page and also appears on Trip Adviser under 'things to do in Ystrad Mynach'.

Although often described as a kind of 'poor man's horse racing' (Cassidy, 2002; Madden, 2010), greyhound racing had wide appeal between the first and second World Wars and audiences in their thousands were not uncommon at some of the larger tracks (Huggins, 1948, p. 102). However, despite attempts to re-package and re-brand the leisure activity to wider audiences in the late 20th and throughout the twenty-first Century, public support has waned, with the last remaining strongholds in areas of high socio-economic depravation. Perhaps this decline is due to the sharp rise in popularity of online forms of gambling (Branigan, 2003, p. 10) coupled with wider public concern for the lives of domesticated non-human animals (Atkinson & Young, 2005, p. 336). Other factors could include recent controversies linking greyhound racing with poor animal welfare outcomes (see, for example, Dearden, 2015; Urwin, 2014).

Political and practical activism for welsh greyhounds

In 2004, Wales found itself the centre of its own high-profile cruelty case. A 'Valleys' kennel assistant was convicted of causing pain and suffering to a young black greyhound called 'Last Hope', who was found alive but mutilated on a rubbish tip in the South Wales Rhymney Valley (BBC, 2004). He had both ears severed, presumably to evade identification through tattooed ear markings which are typically branded onto all race dogs, and had sustained a captive bolt gun wound to the head which had caused severe trauma but not killed him outright.

Greyhound Rescue Wales (hereafter GRW), a greyhound and lurcher rescue and rehoming charity founded in 1993, were at that time pivotal in the community activism which brought the individual responsible for Last Hope's suffering and ultimate death, to justice. In the previous year, the organisation had co-authored a report commissioned by the Welsh Government, which reported that as many as 600 lurchers (greyhound crosses) and greyhounds were being abandoned in Wales each year once their working/racing lives were over (All Party Group for Animal Welfare, 2003).

GRW has continued in their mission to raise awareness of greyhound and lurcher welfare issues in Wales and in 2018, they celebrate their 25th year anniversary. At the time of writing, they conduct their charitable activities from a non-political or 'neutral' position'. This means that they choose not to actively pursue public ethical and political campaigns surrounding the existence of greyhound racing, focusing instead on improving the image of greyhounds among the general public, saving dogs who would otherwise be euthanised and rehoming ex-racing dogs as companions. In 2018, their stated mission is:

> Greyhound Rescue Wales campaigns endlessly to spread the message that Greyhounds and Lurcher's make wonderful pets and companion dogs. We continue to work with key partners to make sure that all Greyhounds in Wales receive protection in law in the future. Greyhound Rescue Wales is an entirely voluntary organisation so all the funds we raise are put to good use with the lowest possible overheads (GRW, 2018).

Greyhound racing as a leisure pursuit continues to be a politically polarised arena. 'Racing-neutral' groups such as GRW have avoided emphasis on whole-system welfare issues, and focus their limited resources on rehoming unwanted industry dogs as companion animals. Whilst a parallel movement, formed predominantly of 'anti-racing' campaigning and rehoming groups, attempts to galvanise public support to favour shutting down the racing industry. Despite these wide internal debates and their resultant disagreements, both positional dichotomies invest heavily in presenting these dogs as suitable companions; converging on their public messaging that ex-racing greyhounds make good pets

(Sighthoundmad, 2013). Together they present this unified vision of ex-race dogs to the public, thus challenging their historical archetypal image as 'aggressive' racing dogs.

Participant greyhound; liminal dogs in public spaces

Public engagement events, such as the Wales-based 'street collections' (a colloquial term used to describe GRW's annual cycle of a city or town-based fundraising activities – see Figure 1) examined for this research, are important to GRW's charitable endeavours for three main reasons. Firstly, they provide income generational opportunities, raising unrestricted funds to support the charity's activities. In the financial year, 2015–2016, 13% of GRW's income was generated through the physical presence and commitments of volunteers at street collections. Secondly, they present a calendar of social and leisure events for the charity's many hundred volunteers, providing spaces of social mobilisation, community cohesion and reinforcement of the common cause of coming together to help greyhounds in Wales.

Figure 1. A GRW street collection event in Newport, August 2016.

And thirdly, these events represent interfacing 'contact zones' (Harraway, 2008) between humans and dogs and thus are ideal spaces through which to examine these encounters vis-à-vis human narrative and non-human perspectives. For GRW, in addition to providing an advocacy platform and public interface for the charity's work, street collections play an important role in presenting alternative representations of former racing dogs. It is this aspect of street collections that will be the basis for this paper.

At street collections, human volunteers talk to the public and give out flyers about the charity's activities and collect donations via charity-branded tins and buckets. They also offer opportunities for passing visitors to meet and engage with 'real' greyhounds, including those seeking adoption. These dogs are referred to as the 'Red Coat Dogs' due to their distinctive bright red jackets worn to signify the status afforded to many contemporary greyhounds as liminal dogs (see Figure 2). In this context, liminality relates to the transitionary space betwixt and between the cosmological categories (Madden, 2010; Srinivasan, 2013) of racing commodity and companion animal, occupied by contemporary (i.e., 'ex-racing') greyhounds. The implicit liminal status of these dogs is here made explicit by the distinctive attire worn by these 'Red Coat Dogs'. Hurn (2011) suggests that clothing animals may be a demonstration of anthropocentric control exerted by humans in order to disguise the (presumed undesirable) animality of non-human others. By enabling physical encounters with actual dogs, which accompany narrative and literature, GRW attempt to demonstrate that dominant mythologies which suggest these dogs to be 'fierce racers' are inaccurate. Furthermore, GRW utilises the opportunities presented through human encounters with 'Red Coat Dogs' to present ex-racing greyhounds as dogs possessing qualities consistent with the image of 'desirable pet'. Dissemination of these messages is a critical objective of street collection events, as is the subsequent rehoming opportunities for dogs that are produced as a consequence.

But who asked the greyhounds? Post-humanist approaches to multispecies ethnographies

The Animal Turn in anthropology and the social sciences has paved the way for an applied anthrozoology which brings non-human animal perspectives into research design and impact (see, for example, Hurn, 2012). While horse racing (e.g. Cassidy, 2002), sled-dog racing (e.g. Kuhl, 2011) and hunting with dogs (e.g. Marvin, 2005) have been subjected to academic scrutiny, the life experiences of racing greyhounds represent a more limited area of research (Dashper, 2017, p. 208). Previous empirical studies exploring the 'greyhound imaginary' (Madden, 2010, p. 504) have reported anthropocentric outcomes associated with greyhound racing as sport and leisure pursuits (Atkinson & Young, 2005, 2008; Huggins, 1948; Madden, 2010), whereas others have examined human experiences associated with rehoming and adoption of former racing dogs (see, for example, Elliot, Toribio, & Wigney, 2010; Howell, Mongillo, Giacomini, Marinelli 2018; Huggins, 1948; Madden, 2010; Thomas, Adams, & Farnworth, 2017). However, all of these studies have prioritised the human voice and narrative over the non-humans in the studies and little attempt has been made to consider the greyhounds' perspectives. Furthermore, no research has been conducted to date on the use of animals in canine-human fundraising street-based events. Consequently, these events represent previously unexamined leisure contexts within which multispecies intra-actions (Barad, 2007) occur between humans and non-human animals. The term 'intra-action' purposely predominates over 'interactions' here in opposition to Cartesian ontologies of individualism and separateness, which populate normative ideas about how 'individual' humans and other animals behave towards each other. Furthermore, for Barad, the meaning [in the encounter] is in the materialising which is brought forth by 'intra-actions', which unsettles individuality as being about a uniform set of 'constituted agents or entities as well as times and places' (Kleinman, 2012). Thinking about 'intra-actions' here allows us to think more expansively about how differences are negotiated in encounters between humans and other animals. This prioritisation of the complexities inherent in how individual beings intra-act is relevant to the post-humanistic approach taken to the multispecies participants in this study, where I have situated racing greyhound participants as individual subjects whose experiences are valued and represented. This is important in order to challenge greyhounds' normative objectified and commodified status as aggregate bodies of the racing industry (Atkinson &

Figure 2. A 'Red Coat Dog' at Bridgend street collection, June 2016.

Young, 2005, 2008; Madden, 2010), and to add to the growing body of literature which situates non-human others as central to explorations of their lives (see, for example, Hurn, 2012, 2018; Madden, 2014; Salazar Parrenas, 2012; Savage-Rumbaugh, 2007). Inspired by the work of relationship commentators including Milton (2005), Bekoff (2001, 2004) and King (2013), I will endeavour to offer some interpretation and understanding of greyhound conspecific interactions, whilst also reflecting on the inherent difficulties in achieving this (that is, I am no greyhound!). This research is an immersive ethnographic and autoethnographic endeavour, which formed one part of the dissertation for my Master's degree in Anthrozoology, submitted in 2016.

Methodology

Having been involved with GRW for over 4 years; initially as a volunteer and more recently, in a paid capacity to provide behaviour advice and support to adopted dogs and their adopters, I occupy what

Hodkinson (2002) describes as a 'relatively unambiguous position as a long term genuine participant' (p. 4). Four GRW street collections facilitated by volunteer humans and their attendant ex-racing greyhounds were attended in May and June 2016. GRW street collection one-day events happen on most weekends in spring and summer from 10am until 4pm and the events I chose to attend represent a diversity of South Wales locations, in terms of population and socio-economics.

Encounters of most anthrozoological interest have been explored which include those that can be defined as multispecies in addition to conspecific interactions:

(1) Greyhound – human volunteer
(2) Greyhound – human member of the public
(3) Greyhound – greyhound

Additionally, I engaged in unstructured discussions through a combination of Skype and face-to-face discussions, with various human individuals to build a broader picture of the motivations, complexities and nuances inherent in bringing together humans and non-human animals on the street, in the name of charity:

(1) Four GRW adopters (contacted via an open Facebook group for the charity's supporters) who adopted their first greyhound after encounters at street collections.
(2) Two individuals representing other dog rescue organisations in Wales; Hope Rescue and Lizzie's Barn were deliberately selected these to be dichotomous in both their public engagement approaches and operations[1].

Scheduled interviews commenced through open-ended questions and developed into detailed discussions. My objective was to reflexively pay attention to all possible data by soliciting my informants' thoughts, feelings, experiences and perspectives in as unstructured a way as possible. I felt this was particularly important in order to mitigate some of my own personal biases, and my pre-existing perspectives on these vis-à-vis greyhound welfare. As a social researcher wishing to delve deep into the workings of greyhound rescue street collections, I have strived to become a 'critical insider' (Hodkinson, 2002, p. 6). Utilising a multi-method ethnographic approach including participant observation, auto-ethnography, discussions and open-ended questions, I have populated disparate spaces in my research methodology. Whilst I concur with Ingold's (2014) assertion that they are far from mutually exclusive (p. 387), attention to the fact that I am simultaneously 'participant' and 'observer' has enabled me to be immersed in the cultural chatter and flow of street collections. With this in mind, I have strived to carry out my research with professional integrity and a 'loving eye' (Acampora, 2005, p. 84) towards all of my informants, both human and greyhound.

Recognising and bringing in the experience of greyhound participants on street collections has been key to my research design and I have sought to utilise experience and ethnography to support this methodology. In particular, I drew on my own 'egomorphic' (Milton, 2005) understandings of greyhounds in order to explore their encounters with the dynamic environment of street collections; recording examples of their interactions with their conspecifics, their familiar caregivers and unfamiliar members of the public. Furthermore, I chose to examine the 'Red Coat Dogs'; those liminal greyhounds wearing red 'jackets' which proclaim their status as dogs-in-need-of-homes.

As an anthrozoologist, I have been faced with a unique opportunity to explore these gatherings, drawing on discourse and narratives residing outside of dominant anthropocentricities. Anthropological advocacy and 'speaking for' an other has been academically problematised, due to complexities inherent in recognizing and choosing from the existence of many kinds of 'others' (Hastrup, Elsass, Grillo, Mathiesen, & Paine, 1990, p. 304). However, the fallacy that as researchers we can (and should) remain within 'neutral objectivity' (Kellet, 2009, p. 23) and unmoved to enact change through our experiences, has equally been exposed. I have chosen to follow Milton's (2005) blueprint for a perception-focused intersubjective relating (p. 265), infused with Bekoff's determination of 'dogomorphism' (2004, p. 495) to bring attentive awareness to the embodied experiences of both my human and greyhound participants (Hammington, 2008; Warkentin, 2010). Indeed, this research is a distinct attempt to bring in the greyhound other – perhaps for the first time – as co-actor, participant and creator of the shared space on the street.

Affect and meaning in the encounter

In my study, intra-actions (Barad, 2007) between ex-racing greyhounds and humans on the street could be described as either economically, emotionally or morally driven. For many of the humans involved in the study, 'on the street' intra-actions included a physical encounter with one or more greyhound participants, producing an affective state, or *auric imprint*.

I coined the term *'auric imprint'* to conceptualise the pervasive effect that these encounters had on some human participants, going beyond the moment of the encounter and prompting a further call to action. This is relevant to generating understandings of street collections as both economically, emotionally and morally relevant spaces of insight for more-than-human leisure experiences. For some human participants in the study, the results of this state compelled them to engage in further moral, economic or emotional exchanges. One participant described how she had felt compelled to economic action after meeting dogs and humans at one event and signed up as a fundraising volunteer for the charity. And across the four street collection events I attended for this research, several people were keen to share their emotional adoption stories after visceral encounters with greyhounds at these events. For these individuals, encounters with greyhounds were 'sticky' memories that had resulted in their continued commitment to the charity and its work. At one event, I spoke at length with one of the new street collection co-ordinators about his personal journey through volunteering with the charity. He reflected on how greyhounds had 'gotten under his skin', and laughed to himself at how quickly this had occurred. When I asked how this felt for him, he replied that it (volunteering for greyhounds) felt 'in his heart' the right thing to do. The sense of being physically mobilised by one's heartfelt purpose is an experience familiar to me. I have often meditated on how things have shifted and aligned in my own life to bring me to this place where I am immersed in a quest to better understand the lives of this population of dogs and to add to the emerging body of work which is finding ways of speaking with and for non-human animals.

Certain encounters brought people into a mindful presence with their own memories, experiences and thoughts of their own companion dogs. Recounting one interaction between a member of the passing public and a greyhound, I observed as Ben was physically manipulated and held in place on a tight lead by his human caregiver, in order to be stroked by the attendant unfamiliar human who had remarked that Ben reminded her so much of her own dog. Ben's human caregiver responded reciprocally and spoke of how;

'Ben loves street collections. He loves meeting new people'.

Ben appeared to passively comply, offering no obvious resistance to the way in which the manoeuvrer of the lead held his body into place in order to be stroked. However, I could see the familiar light puffing of cheeks and quick darting of eyes that, through close observation and study, I have come to understand as demonstrations of inner anxiety in some of these dogs. (Dawson, 2018; Heath, 2018). At times, I found it a challenge not to modify the interactions; feeling the prickly push-pull of my role as sensitive ethnographer (Sanders & Arluke, 1993, p. 358) vis-à-vis my natural drive to want to interpret, adapt and at times, protect and offer sanctity to my greyhound friends. Whilst the human carers of the greyhounds at street collections cared deeply about the cause of rescuing and rehoming greyhounds dogs in their care, I suggest that there is a certain cognitive dissonance around how 'real' individual dogs may experience these events.

Tami and Gallagher (2009) discovered that practical experience alone of being around dogs, is an insufficient predictor of the correct reading of dog behaviour (p. 168), and research from the field of Animal Assisted Interventions suggests that dog guardians are poor at both noticing and interpreting their dog's body language and communication (see, for example, Glenk et al., 2014, p. 105; Hatch, 2007, p. 40; Horrowitz, 2009, p. 451). In addition, humans may misdiagnose dog behaviour (see, for example, Payne, Bennett, & McGreevy, 2015, p. 73) and miss subtle canine stress (see, for example, Mariti et al., 2012, p. 217). The implications of this include a muddying of understandings about canine agency and the common conflation of consent with a lack of demonstrative protest, which is problematic to the well-being of individual greyhounds at public events.

Gluck's (2010) work, reflecting on his own journey from research psychologist to an ethical commentator, provided inspiration and a space through which to explore my ambivalence about the human interpretation of greyhound consent to be touched. He warned of the insidious nature of growing indifference (p. 122), suggesting that we must work hard to keep the otherness and identity of individual animals alive (p. 124) so as to allow us to retain sensitivity and a sense of morality. Although Gluck's work related to the use of animals in laboratory experiments, the parallels with 'using' greyhounds at street collections resonate for me. On the few occasions that I did ask questions and pass comment on the well-being of a certain greyhound, I felt a wall of defensiveness and the subsequent withdrawal of my human participants. The subsequent splitting off of what one sees from what one feels can result in what Arluke (2004) describes as 'play acting it' (p. 199), which can result in 'a lingering moral stress' (p. 201), which is certainly something I can relate to. Ever mindful of this, I worked hard to cultivate distance where necessary during this research in order to discern a more critical form of analysis (Hodkinson, 2002, p. 5).

Implicit and explicit narratives at street collections

The topic of greyhound racing was largely absent in the narratives at the street collections attended for this research. Greyhound racing is a largely unknown activity, by all except those directly involved. In their work exploring sport-related violence and non-human animals, Atkinson & Young (2005, 2008) describe what they term a 'greyhound racing figuration'; a system of overlapping 'power players' situated within an animal-sport complex. For them, this includes the various denominations of greyhound welfare organisations, who they suggest are 'complicit in the acts of abuse through conscious acquiescence' (2005, p. 352). Furthermore, evasion of the challenging political issues associated with 'the total greyhound imaginary' (Madden, 2010, p. 504), leaves a conspicuous blank space in what Madden describes as 'all manifestations of the ways human think about, write about and picture greyhounds' (p. 504).

In the previous research, I have problematised the invisibility of the greyhound racing industry and how this limits attempts to understand the lives of racing greyhounds (Sands, 2015). I have drawn parallels with Pachirat's (2011) ethnography of the silo working of a modern slaughter-house where the smooth operations of the 'dirty work' (Sanders, 2010) of killing animals relies on an intricate and ordered system of knowing, not knowing and, in some cases perhaps, not wishing to know. I suggest that Atkinson and Young's (2005) 'greyhound racing figuration' (p. 352) also foregrounds the enactment of a similar Foucauldian system of power and control. The intricate systems or 'power players' within this figuration coalesce to enable the facilitation of animal abuses in the publically concealed 'back regions' (Goffman, 1959) of greyhound racing operations.

Greyhounds are domesticated dogs. Yet as artefacts of the greyhound racing industry, they occupy a liminal space, seemingly 'slip[ping] between the cracks' (Srinivasan, 2013, p. 6) and 'straddling the [cosmological] categories' (Madden, 2010, p. 505) ascribed to working (Cassidy, 2002; Marvin, 2005) and companion dogs. Racing greyhounds are difficult to know and the racing industry's systematic and widespread concealment of the husbandry and socialisation practices of these dogs have resulted in a population of domestic dogs whose lives are largely without public scrutiny. Street collection events evoke a bringing to mind of 'real' greyhound lives through actual physical encounters with a favoured few ex-racing dogs. Yet as greyhound 'racing neutral' groups choose to bypass meaningful public engagement on whole welfare debates, these events become situated as representations of 'regulation players' within Atkinson and Young's (2005) figuration.

By refusing to publically acknowledge the systematic challenges caused to these dogs through the greyhound racing production system, street collection events fail to make explicit what may be implicitly present in the ancestral lives of the attendant dogs. Atkinson and Young (2008) discuss the networks of pain and trauma that perpetuate this status quo. They write; 'As long as the demands and effects of competition (anxiety, pain, injury) are hidden from spectators, greyhound racing is not problematized as victim-producing and thus maintains it's acceptable veneer'. (p. 85) Such realities exist only in the unknowable and unspoken 'back region' (Goffman, 1959) terrains of street collections. Therefore, events organised by 'racing-neutral' groups may miss opportunities to advocate for the unseen members of this vulnerable population of dogs, in whom they invest so deeply.

Human caregivers and representations of greyhounds

For the greyhound individuals in this study, their responses within and between street collection events were diverse and individualised. Although all dogs featured in this study were categorised as, 'ex-racing greyhound', each participant demonstrated an array of behaviours and responses suggestive of their subjective agency and consciousness vis-à-vis the street collection environment.

King (2013) reminds us that 'we should not make universality a criterion for the existence of a phenomenon' (p. 29). At one event I attended, a member of the public rode a bicycle speedily past at close quarters to where the dogs were resting on their patchwork of makeshift beds. In response to this activity, one dog stood and barked in rapid succession in a display one could perceive to be an alarm, another tried to give chase, although, impeded by the lead, was flung up into the air, landing down onto the concrete with a thud. Another still tried to escape; becoming entangled in her lead and yelping in panic, whilst two of the participant greyhounds offered no outward response at all. Some of the same dogs attended multiple events, although they could have been considered different dogs, such was the change in their behaviour between one event and the next. On one occasion, one street collection event co-ordinator expressed surprise at the change he had noted in one greyhound's demeanour:

'Is that Katie? I would never have known! She never comes to the front like that. She's finally coming out of herself, then?'

Static understandings about the fixed nature of dog behavioural traits currently reside within the companion dog field (Payne et al., 2015, p. 73). The broad brush that is 'animal temperament' remains the most utilised yardstick with which to measure the volition of individual dogs (Clothier, 2002, p. 103). Oftentimes, the behaviour of known animals is conceived as fixed and largely resistant to the daily variety of environmental stimuli, which would explain why a change in greyhound Katie's apparent demeanour (above) was significant enough to warrant comment. However, as I have explored in the previous research (Sands, 2015), by simply asking more questions about the multiplicity of non-human animal experience, we are able to begin to acknowledge the rich assortment of their inner worlds.

In addition to thinking of them as a homogenous group, human caregivers may place greyhounds in a liminal space; as a homogenous collective whose caricatured 'couch potato' image makes them ideal for a static meet and greet events. A number of my human participants described their dogs affectionately as being 'typical greyhounds'. One commented;

'Yes he's exactly like they tell you greyhounds are... a typical couch potato. Sleeps most of the time.'

In lay discourse, greyhounds are also often portrayed as comical artefacts, and 'not-like-other-dogs'; 40 mph 'couch potatoes' who love nothing more than sleeping all day. Academic studies featuring detailed behavioural analysis of ex-racing greyhounds as a discrete population of dogs are limited, although one study suggested that greyhounds tend to be less vocal than other breeds (Elliot et al., 2010, p. 131) and another listed them as least likely to show stranger-directed aggression towards humans (Duffy, Hsu. & Serpell, 2008, p. 454). However, a predominant breed trait of greyhounds is their passive coping style in response to stressors in their environment (Wormald, Lawrence, Carter, & Fisher, 2016), which may make their inhibited anxiety responses easy to mistake for calmness and contentment. Greyhounds are a less common breed of companion dog choice and the vast majority that are found in family homes in the UK are the unwanted dogs of the racing dog industry. This means that they will have undergone the breeding, socialising and training needed to produce dogs who will chase moving objects. In direct contrast to their dominant image as 'perfect pets', as greater numbers of these dogs enter companion homes, a potentially concerning trend may be emerging. In one study of nearly 8,000 behaviour referrals to an Australian vet clinic, greyhounds were overrepresented in the list of breeds referred for aggression, barking and anxious behaviour (Col, Day, & Phillips, 2016, p. 9), overtaken only by guarding and herding breeds. And a 2018 study of ex-racers in companion homes in Italy cited predatory behaviour towards other animals as a concern for adopters in nearly half of the dogs in the study (Howell et al., 2018). In addition, there is a growing groundswell of concerned advocates, dog trainers and veterinary behaviour professionals in both the UK and Australia speaking out about the gaps, myths and half-truths in dominant public discourse about these dogs (see, for example, Eddie, 2018; Fawcett, 2017; Forbes, 2018; Sands, 2018).

Developing greyhound perspectives at anthropocentric leisure events

Applied anthrozoological discourse suggests that as a researcher, one must engage in 'conscious partiality' (Sanders & Arluke, 1993, p. 358), becoming a sensitive ethnographer, who at least attempts to 'raise the context awareness' (Hastrup et al., 1990, p. 306) of the places and spaces that humans and animals co-inhabit.

Multispecies interactions within leisure spaces are beginning to attract wider academic interest (see, for example, Carr, 2014, 2015; Dashper, 2017; Young & Carr, 2018). Birke's (2009) paper provides both focus and inspiration for asking the kind of questions that seek to address non-human animal experience in anthropocentric spaces. Hatch (2007) and Zamir (2006) draw our attention to the perspective of the non-human animal in therapeutic settings, reminding us that, for some animals, such encounters may be individually and ethically problematic.

King (2013) considered short-term positive association to be a suitable criterion for friendship, and Townley (2010) suggests that friendship 'requires mutuality, but not symmetry' (p. 48). Furthermore, in a year-long study of the affiliative and agonistic behaviours of day care and non-kin familial dogs, Trisko, Sandel, and Smuts (2016), suggested that dogs have highly differentiated relationships with one another, though the authors also note the general absence of academic literature specifically exploring intraspecific 'friendly behaviour' in companion dogs (p. 3).

I witnessed several intimate exchanges and 'gentle affiliative behaviours' (Trisko et al., 2016) including coat licking, head resting and nibbling between two or more greyhounds, which could have been missed by the inattentive human eye. These encounters reflected the 'mutuality without symmetry' suggested by Townley (2010) as indicative of friendship, suggesting that, in stark contrast to their representations as predictable 'animatronic dogs', greyhounds are intentional, emotionally responsive beings (Bekoff, 2001; Milton, 2005, p. 257). One intimate exchange between 'Red Coat Dog' (Jaffa) and new greyhound arrival (Kelly) at one event was particularly illustrative of this:

As Kelly approached I watched Jaffa, poised in sphinx position with head resting on his front paws, following her movement with intense eyes. Without meeting his gaze, Kelly moved lightly and fluidly, as if on her toes, towards the empty bed space next to where Jaffa was laying. Turning once in a circle, she came resting down, and into a tight fox-like curl. Still, Jaffa remained motionless; the only clue to his minded subjectivity of Kelly's felt presence remained the animation of his eyes, which now offered occasional dancing sideways glances to where she was situated. Then, after several moments of space and inactivity, Kelly part-uncurled herself. Craning her neck and head, she came to rest these gently across Jaffa's immobile shoulders. I sensed a moment of tension in the encounter, noting Jaffa's momentary twitching of shoulders and the sharp sideways swipe of his eyes, before noting Kelly's audible extended deep breath, and sigh. I watched, all together captivated and moved, as Jaffa's eyes started to flicker, soften, and then momentarily close. They rested like this for several minutes.

For this research, I was keen to engage my human informants in discussions about their perceptions of these kinds of interactions, to stimulate wider thinking about their greyhound's experience of street collections. As 'owned' domestic dogs, greyhound participants were almost entirely reliant upon their human partners and guardians at these events. Although academically critiqued and problematised (see, for example, Hastrup et al., 1990, p. 304; Kellett, 2009; p. 23; Scheper-Hughes, 1995, p. 415) human activities at street collections could be understood as anthropological advocacy. However, some of the human caregivers I interviewed seemed challenged by the competing positions and priorities presented by this advocacy for greyhounds. There was a sense of both wanting to be able to represent the charity and its core purpose but also an ambivalent awareness that not all dogs enjoyed the experience. As one volunteer revealed to me at the end of a long discussion about the role of the dogs at street collection events;

> I say that Jude loves coming on street collections. But I'm not honestly sure that she does. I think sometimes that it's me who enjoys coming to these things… and maybe she would prefer to just stay at home. But it's not the same without her, is it? And the public love to see the dogs.

For some of my participants, the weight of responsibility to the cause of rescuing greyhounds drives their commitment and, in some cases, muddies the waters of individual greyhound well-being. I suggest that the generic greyhound rehoming message of the past 30 years, which culturally transmits well-intentioned, yet loaded memes of 'perfect pets' and '40 mph couch potatoes', has directly contributed to the practical challenges volunteers face in recognising and acting upon individual greyhound subjectivities at street collections.

As a greyhound advocate and applied anthrozoologist, I was keen to ensure that my greyhound participants' perspectives were heard through this research. Milton (2005) proposed that we should look to our personal experiences and connections with individual non-human animals in order to help shape our determination of them; an approach she referred to as 'egomorphism' (p. 261). Bekoff (2004) has applied this to the ethological study of dogs, suggesting that one needs to inhabit the perspective of 'dogomorphism' to truly understand dog experience (p. 495). By utilising these approaches and asking questions about the multiplicity of greyhound experience and inner worlds, I was able to provoke changes in human narratives by some caregivers. At one event, I supported a regular volunteer to unpack her inner confusion about whether she was rewarding her greyhound's fear by comforting him when he recoiled from the touch of unfamiliar men at street collections. Seeking to avoid discussions about the technical accuracies of whether it was even possible to reinforce an emotion I instead offered an empathic ear, allowing space for her internal conflict and intuitive doubt. We began by noticing the external environmental conditions that made this particular event challenging. It was particularly cold on this day and many of the dogs, including Parker, were huddled together with other dogs. I then encouraged her to quietly focus on Parker for a few moments, attending closely to both her embodied knowledge (Hammington, 2008) of Parker and his observed embodied movements (Warkentin, 2010). Together, we explored Parker's outward demonstrations of his inner state; his constant wide pant, quickly darting eyes and the placement of his body at the back of the gathering, shielded by the other dogs. This attentive awareness exercise seemed to stimulate reflections on other aspects of Parker-as-person. These included his likes and dislikes, his propensity to 'be the wallflower' rather than 'party animal' and his generalised fear of the unknown, which his caregiver attributed to his challenging former racing life. After this discussion, during which I encouraged the exploration of Parker's perspective utilising both 'dogomorphic' and ethological approaches (Bekoff, 2007), Parker's caregiver started to consider what it may feel like 'to be [her greyhound]' (Madden, 2014, p. 285). The change to her perspective through this process was profound, and she later recounted to me that she could now see how Parker did not enjoy the experience of street collections and so she would not be taking him again.

Conclusion

In this paper, I have focused on examining human-canine intra-actions through encounters that occur on the street between humans and greyhounds at street fundraising events. I have offered commentary on conspecific interactions that I observed at a selection of these events and posed questions about non-human agency in anthropocentric spaces; exploring how ex-racing greyhound subjectivities converge and collide with human narrative and dominant discourse about these dogs.

During this research, asking questions about the multiplicity of greyhound experience and inner worlds provoked changes in human narratives by some caregivers. Thus, street collections may provide spaces of insight where human behaviour change for non-human animals can be tested and enacted, although more work is needed to determine and refine the levers for this.

Moving public discourse away from the limiting representation that is the homogenous greyhound, will bring us towards an egomorphic understanding of the diversity with which non-human animals express their own aliveness. To this end, future research may develop meaningful methodologies that enable greyhounds, charity volunteers, caregivers, and the general public to encounter each other in mutually beneficial ways.

Couch potato, you have had your chips! The time has come for an individualised and subjective greyhound to emerge. All greyhound names have been changed to protect their identities and to preserve the confidentiality of their human caregivers.

Note

1. Hope Rescue, a South Wales-based charity, is a co-ordinating rescue who save dogs from emergency situations in Local Authority-run kennel facilities. As an organisation, they are particularly public-facing and rely heavily on the support of foster homes, to help rehabilitate some of the dogs they take in. Lizzie's Barn is a small independently family-run rescue who operate a private sanctuary facility for hard-to-home dogs. Both of these contacts were made via a former colleague, who also works in animal rescue.

Disclosure statement

No potential conflict of interest was reported by the author.

ORCID

Kerry L. Sands iD http://orcid.org/0000-0001-6031-3744

References

Acampora, R. (2005). Zoos and eyes: Contesting captivity and seeking successor practices. *Society & Animals*, *13*(1), 69–88.

All Party Group for Animal Welfare. (2003). *The fate of racing greyhounds and working lurchers in Wales*. Wales, UK: National Assembly Wales.

Arluke, A. (2004). The use of dogs in medical and veterinary training: Understanding and approaching student uneasiness. *Journal of Applied Animal Welfare Science*, *7*(3), 197–204.

Atkinson, M., & Young, K. (2005). Reservoir dogs: Greyhound racing, mimesis and sports-related violence. *International Review for the Sociology of Sport*, *40*(3), 335–356.

Atkinson, M., & Young, K. (2008). *Deviance and social control in sport*. UK: Human Kinetics.

Barad, K. (2007). *Meeting the Universe Halfway: Quantum physics and the entanglement of matter and meaning.* Durham: Duke University Press.

BBC (2004, November 22). Man guilty of mutilating dog. Retrieved from http://news.bbc.co.uk/1/hi/wales/south_east/4032981.stm

Bekoff, M. (2001). The evolution of animal play, emotions, and social morality: On science, theology, spirituality, personhood, and love. *Zygon, 36*(4), 615–655.

Bekoff, M. (2004). Wild Justice and fair play: Cooperation, forgiveness and morality in animals. *Biology and Philosophy, 19*, 489–520.

Bekoff, M. (2007). *The emotional lives of animals.* New York: New World Library.

Birke, L. (2009). Naming names – or, what's in it for the animals? *Humanimalia, 1*(1), 1–9.

Branigan, C. A. (2003). *Adopting the racing greyhound* (3rd ed.). Hoboken, NJ: W. Publishing.

Carr, N. (2014). *Dogs in the leisure experience.* Wallingford: CABI.

Carr, N. (ed). (2015). *Domestic animals and leisure.* Basingstoke, UK: Palgrave Macmillan.

Cassidy, R. (2002). *The sport of kings.* Cambridge: Cambridge University Press.

Clothier, S. (2002). *Bones would rain from the sky: Deepening our relationships with dogs.* New York, NY: Warner Books.

Col, R., Day, C., & Phillips, C. J. C. (2016). An epidemiological analysis of dog behavior problems presented to an Australian behavior clinic, with associated risk factors. *Journal of Veterinary Behavior, 15*(16), 1–11.

Dashper, K. (2017). *Human-animal relationships in equestrian sport and leisure.* Abingdon: Routledge.

Dawson, K. (2018, July 5). *Personal communication.* Sydney Australia: Veterinary Behaviourist.

Dearden, L. (2015, February 17). Live piglets and possums filmed being used to illegally bait racing greyhounds in Australia. Retrieved from http://www.independent.co.uk/news/world/australasia/live-piglets-and-possums-filmed-being-used-to-illegally-bait-racing-greyhounds-in-australia-10050585.html

Duffy, D. L., Hsu, Y., & Serpell, J. A. (2008). Breed differences in canine aggression. *Applied Animal Behaviour Science, 114*, 441–460.

Eddie, R. (2018, September 20). Third of retired greyhounds could be put down for failing rehoming test. Retrieved from https://thenewdaily.com.au/news/state/nsw/2018/09/20/greyhound-racing-nsw-rehoming/

Elliot, R., Toribio, J. L. M. L., & Wigney, D. (2010). The Greyhound Adoption Program (GAP) in Australia and New Zealand: A survey of owners' experiences with their greyhounds one month after adoption. *Applied Animal Behaviour Science, 124*, 121–135.

Fawcett, A. (2017, March 15). Rehoming retired greyhounds – The challenges. Retrieved from http://www.smallanimaltalk.com/2017/03/rehoming-retired-greyhounds-challenges.html

Forbes, T. (2018, July 15). Greyhounds 'docile and low maintenance' but vets warns of adoption risk for dogs bred to race. Retrieved from http://www.abc.net.au/news/2018-07-14/animal-behaviour-expert-warns-of-adopted-greyhound-bite-risk/9950038

Glenk, L. M., Kothgassner, D. O., Stetina, B. U., Palme, R., Kepplinger, B., & Baran, H. (2014). Salivory cortisol and behaviour in therapy dogs during animal assisted interventions: A pilot study. *Journal of Veterinary Behaviour, 9*, 98–106.

Gluck, J. P. (2010). Discovering the way back to the solid ground of ethical uncertainty: From animal use to animal protection. In G. A. Bradshaw & N. Cater (Eds.), *Minding the animal psyche. Spring; A journal of archetype and culture* (Vol. 83, pp. 119–133). New Orleans, Louisiana: Spring.

Goffman, E. (1959). *The presentation of self in everyday life.* Garden City, NY: Doubleday-Anchor.

Greyhound Rescue Wales. (2018). About us. Retrieved from https://greyhoundrescuewales.co.uk/about-greyhound-rescue-wales/

Hamington, M. (2008). Learning ethics from our relationships with animals: Moral imagination. *International Journal of Applied Philosophy, 22*(2), 177–188.

Haraway, D. (2008). Training in the contact zone: Power, play, and invention in the sport of agility. In B. Da Costa & K. Philip (Eds.), *Tactical biopolitics: Art, activism, and technoscience* (pp. 445–465). Cambridge, MA: MIT press.

Hastrup, K., Elsass, P., Grillo, R., Mathiesen, P., & Paine, R. (1990). Anthropological advocacy: A contradiction in terms? *Current Anthropology, 31*(3), 301–311.

Hatch, A. (2007). The view from all fours: From the animals' perspective. *Anthrozoös, 20*(1), 37–50.

Heath, S. (2018, July 18). *Personal communication.* Chester, UK: Veterinary Behaviourist.

Hodkinson, P. (2002). *Goth: Identity, style and subculture.* London: Bloomsbury Publishing.

Horowitz, A. (2009). Disambiguating the 'guilty look': Salient prompts to a familiar dog behaviour. *Behavioural Processes, 81*, 447–452.

Howell, T. J., Mongillo, P., Giacomini, G., & Marinelli, L. (2018). A survey of undesirable behaviors expressed by ex-racing greyhounds adopted in Italy. *Journal of Veterinary Behavior, 27*, 15–22.

Huggins, M. (1948). 'Everybody's going to the dogs'? The middle classes and greyhound racing in Britain between the wars. *Journal of Sport History, 34*(1), 97–120.

Hurn, S. (2011). Dressing down: Clothing animals, disguising animality? *Civilisations, 59*(2), 109–124.

Hurn, S. (2012). *Humans and other animals: Cross cultural perspectives on human-animal interactions.* London: Pluto Press.

Hurn, S. (2018). Encounters with dogs as an exercise in analysing multi-species ethnography. In J. Lewis (Ed.), *SAGE research methods dataset* (pp. 1–17). London: Sage.

Ingold, T. (2014). That's enough about ethnography! *HAU: Journal of Ethnographic Theory, 4*(1), 383–395.

Kellett, P. (2009). Advocacy in anthropology: Active engagement or passive scholarship? *Durham Anthropology Journal, 16*(1), 22–31.

King, B. J. (2013). *How animals grieve.* US: University of Chicago Press.

Kleinman, A. (2012). Intra-actions. *Mousse, 34*, 76–81.

Kuhl, G. (2011). Human-sled dog relations: What can we learn from the stories and experiences of mushers? *Society & Animals, 19*, 22–37.

Madden, R. (2010). Imagining the greyhound: 'Racing' and 'rescue' narratives in a human and dog relationship. *Continuum: Journal of Media and Cultural Studies, 24*(4), 503–515.

Madden, R. (2014). Animals and the limits of ethnography. *Anthrozoös, 27*(2), 279–293.

Mariti, C., Gazzano, A., Lansdown Moore, J., Baragli, J., Chelli, L., & Sighieri, C. (2012). Perception of dogs' stress by their owners. *Journal of Veterinary Behavior, 7*, 213–219.

Marvin, G. (2005). Disciplined affections: The making of an english pack of foxhounds. In J. Knight (Ed.), *Animals in person: Cultural perspectives on human-animal intimacies* (pp. 61–77). Oxford: Berg.

Milton, K. (2005). Anthropomorphism or egomorphism? The perception of non human persons by human ones. In J. Knight (Ed.), *Animals in person: Cultural perspectives on human-animal intimacies* (pp. 255–271). Oxford: Berg.

Pachirat, T. (2011). *Industrialised slaughter and the politics of sight.* New Haven and London: Yale University Press.

Payne, E., Bennett, P. C., & McGreevy, P. D. (2015). Current perspectives on attachment and bonding in the dog-human dyad. *Psychology Research and Behavior Management, 8*(p), 71–79.

Salazar Parrenas, R. J. (2012). Producing affect: Transnational volunteerism in a Malaysian orangutan rehabilitation centre. *American Ethnologist, 39*(4), 673–687.

Sanders, C. R. (2010). Working out back: The veterinary technician and 'Dirty Work'. *Journal of Contemporary Ethnography, 39*(3), 243–272.

Sanders, C. R., & Arluke, A. (1993). If lions could speak: Investigating the animal-human relationship and the perspectives of nonhuman others. *The Sociological Quarterly, 34*(3), 377–390.

Sands, K. (2015) *Greyhound Mythology: How do dominant ideologies of breed characteristics for greyhounds influence their transition in life from racetrack to companion animal?* (Unpublished Applied Anthrozoology module essay), University of Exeter, Exeter.

Sands, K. (2018, March 21). Easy pets or fast dogs? The problem with labelling greyhounds. Retrieved from: https://theconversation.com/easy-pets-or-fast-dogs-the-problem-with-labelling-greyhounds-93525

Savage-Rumbaugh, S. (2007). Welfare of apes in captive environments: Comments on, and by a specific group of apes. *Journal of Applied Animal Welfare Science, 10*(1), 7–19.

Scheper-Hughes, N. (1995). The primacy of the ethical: Propositions for a militant anthropology. *Current Anthropology, 36*(3), 409–440.

Sighthoundmad. (2013). *Behind the lights, the tote and the non-starters.* UK: Grey.

Srinivasan, K. (2013). The biopolitics of animal being and welfare: Dog control and care in the UK and India. *Transactions of the Institute of British Geographers, 38*, 106–119.

Tami, G., & Gallagher, A. (2009). Description of the domestic dog (Canis familiaris) by experienced and inexperienced people. *Applied Animal Behaviour Science, 120*, 159–169.

Thomas, J. B., Adams, N. J., & Farnworth, M. J. (2017). Characteristics of ex-racing greyhounds in New Zealand and their impact on re-homing. *Animal Welfare, 26*, 345–354.

Townley, C. (2010). Animals as Friends. *Between the Species, X*, 45–59.

Trisko, R. K., Sandel, A. A., & Smuts, B. (2016). Affiliation, dominance and friendship among companion dogs. *Behaviour, 1*, 1–33.

Urwin, R., (2014, October 26). The mystery of the 1,000 greyhounds who retire and then vanish. Retrieved from http://www.independent.co.uk/news/uk/home-news/the-mystery-of-the-1000-greyhounds-who-retire-and-then-vanish-9818554.html

Warkentin, T. (2010). Interspecies etiquette: An ethics of paying attention to animals. *Ethics and the Environment, 15*(1), 101–121.

Wormald, D., Lawrence, A. J., Carter, G., & Fisher, A. D. (2016). Physiological stress coping and anxiety in greyhounds displaying inter-dog aggression. *Applied Animal Behaviour Science, 180*, 93–99.

Young, J., & Carr, N. (2018). *Domestic animals, humans, and leisure: Rights, welfare, and wellbeing.* UK: Routledge.

Zamir, T. (2006). The moral basis of animal assisted therapy. *Society and Animals, 14*(2), 179–199.

What's in it for the cats?: cat shows as serious leisure from a multispecies perspective

Emily Stone ⓘD

ABSTRACT

The breeding and showing of pedigree cats provides a novel lens through which to explore more-than-human intersections within leisure. Based on multispecies ethnographic fieldwork at multiple cat shows across the United Kingdom and on interviews with those who breed and exhibit cats, this article explores the relevance of the concept of 'serious leisure' to the 'hobby' of cat showing and asks 'what's in it for the cats?' For the human participants, it provides opportunities for social interaction, knowledge and skill progression, as well as contributing to individual and collective identities. There are also substantial costs to this form of engagement for the human, but particularly, I argue, for the feline competitors. This article asserts that human leisure needs are given precedence over the well-being of feline individuals. Exhibition spaces and show requirements present substantial challenges for many of the cats involved and can limit, or even deny altogether, the expression of feline agency. This is not to state that there are not relations based on intersubjectivity between cats and humans in cat showing, but the focus of the activity and its power dynamics result in human interests being prioritised.

Introduction

The breeding and showing of domestic cats (*Felis catus*) provides an underexplored human–feline 'contact zone' (Haraway, 2008b), and a valuable opportunity to examine more-than-human intersections within a leisure context. This 'hobby', known as the 'cat fancy', incorporates those who deliberately breed 'pedigree' cats (cats of specific 'breeds' who are registered and certified with cat fancy organisations) for their exhibition at 'cat shows' as a form of competition that primarily evaluates the cats on their conformation to a set of physical and behavioural traits. Breeders also sell cats as companion animals or to other breeders.

Utilising insights gathered during multispecies ethnographic fieldwork at cat shows[1] and interviews with human cat fanciers across the United Kingdom, this article will investigate the relevance of the concept of 'serious leisure' to the cat fancy, with a focus on the cat show experience and the consequences of this type of engagement for all the actors involved. In general, less mainstream avocations involving domesticated species other than dogs and horses have been neglected within research (Carr, 2015, p. 24). Hirschman (1994) noted that pedigree pet shows had been overlooked ethnographically and this is still predominantly the case, despite them offering a window into our conflicting attitudes and ambiguous relationships with other species. Where studies have been conducted into human-pet leisure activities, they have primarily focused on the human side of the relation (Carr, 2014; e.g. Baldwin & Norris, 1999; Gillespie, Leffler, &

Lerner, 2002; Hultsman, 2012). I will attempt to rectify this omission and extend the concept of 'serious leisure' beyond an anthropocentric frame, centring on the relationships between feline and human lives and highlighting feline subjective experiences in order to address the question, 'what's in it for the animals?' (Birke, 2009, p. 1).

Cat shows as 'serious leisure'

'Serious leisure', a term coined by Stebbins (1982), is defined as the:

> Systematic pursuit of an amateur, hobbyist, or volunteer core activity that people find so substantial, interesting, and fulfilling that, in a typical case, they launch themselves on a (leisure) career centred on acquiring and expressing a combination of special skills, knowledge, and experience (Stebbins, 2015, p. 3).

As noted above, serious leisure participants are generally categorised into three types. These include 'amateurs' who are involved in an activity in which professional counterparts also participate, such as sports or the arts, 'career volunteers' who provide uncoerced help beneficial to others and themselves, and, arguably the most appropriate category to describe many cat fanciers, hobbyists. A hobby is described as 'a specialised pursuit beyond one's occupation, a pursuit one finds particularly interesting and enjoys doing because of its durable benefits' (Stebbins, 1982, p. 260). However, there can also be commercial actors within cat breeding, for example there are those who breed primarily for financial gain, often referred to as 'backyard breeders'. Furthermore, there can be 'voluntary' aspects with participants volunteering to assist in the organisation of shows and learning skills along the way. Stebbins (2015) refers to the activity of dog sports as described by Gillespie et al. (2002) as a 'making and tinkering' hobby, in the same category as crafts, due to the desire from human participants to 'make' their dogs into competitive animals (p.18). Following this line of thought, the exhibiting of purposely bred cats could also be regarded as such a hobby. This would imply that the cats are seen in part as 'objects' to be shaped. All in all, it appears that the cat fancy is an example of 'mixed serious leisure' in that a person can partake in multiple serious leisure categories (Stebbins, 2015), suggesting more complexity than the above system of categorisation readily allows.

Stebbins (2015) refers to the study of serious leisure as a form of 'positive sociology' because unlike the predominant 'problem-centred' view of sociology, it is focused on the study of people's organisation of their lives in pursuit of fulfilment (p.xii). This view draws on a long-standing perception of leisure, traceable to Greek and Roman thought, as 'a sought after and inherently positive state for those who were able to engage in it' (Gallant, Arai, & Smale, 2013, p. 95). However, this notion of leisure overlooks the other actors who may be affected such as family and friends (Gallant et al., 2013), or those directly involved in the activity for whom it may not be rewarding, including the other-than-human participants.

Gallant et al. (2013) critique the concept of 'serious leisure' on the grounds that it is almost wholly positive and neglects to report the costs associated with participation. For the humans involved, serious leisure pursuits are often seen as having substantial benefits, including personal rewards such as the development of skills and knowledge, self-gratification, the enhancement of self-image, and a number of social benefits incorporating interaction with others, a feeling of group accomplishment, and group maintenance (Stebbins, 2015, p. 9). There are also costs involved, however, such as participants experiencing interpersonal conflict through power dynamics involved in the activity and its organisation, tension between family and friends outside of the activity and dissatisfaction from bad performances (Lamont, Kennelly, & Moyle, 2014). There is also the possibility of participants taking risks to advance their leisure 'careers' that could result in injury or alienation (Gallant et al., 2013, p. 96).

In general, Stebbins (2015) argues that the rewards from serious leisure are usually perceived as far outweighing any costs. However, like Gallant et al. (2013), I am wary of describing the activity in dichotomous terms, such as 'positive' and 'negative', as this negates the complexity of the experience and the fact that a benefit to one can be at the expense of another. Furthermore, research on serious leisure has often focused on the individual (Gallant et al., 2013), when it also has implications at a group or community level.

The activity of showing pedigree animals can be seen as a subculture in that it has its 'own specialised jargon, rituals of behaviour, and status hierarchies' (Hirschman, 1994, p. 630). These are transmitted socially as members of the group interact (Green, 2001). As such, serious leisure contributes to 'a sense of solidarity with the collective' in addition to the individual (Gallant et al., 2013, p.102; Rojek, 2001). As I will discuss below, however, these group dimensions are a potential source of friction. Moreover, although 'serious leisure' provides a useful framework for exploring the level of commitment devoted to the activity, as will become apparent, the cat fancy can become more than a 'hobby' for many participants and instead becomes 'a way of life', in part due to the quotidian care requirements that make it difficult to compartmentalise the activity (Dashper, 2017, p. 3). To account for some of the issues associated with the concept of 'serious leisure', Gallant et al. (2013) propose a revised definition as:

> The committed pursuit of a core leisure experience that is substantial, interesting, and fulfilling, and where engagement is characterised by unique identities and leads to a variety of outcomes for the person, social world, and communities within which the person is immersed (p.104).

This definition is less prescriptive and allows for more nuance within people's experiences. Yet, the other-than-human dimension is still overlooked. My human condition, of course, prevents me from truly knowing what it is to be a cat, and from fully understanding what matters to them (Lloro-Bidart, 2018; Warkentin, 2010). However, I believe that as researchers we have the ability and the responsibility to attempt to portray their experiences as well (see Birke, 2009; Gruen, 2015; Hurn, 2018; Milton, 2005). By doing so, I will incorporate the feline experience and highlight how cat showing practices can result in human interests being prioritised, detrimentally impacting upon feline well-being.

Methods

Between April 2017 to April 2018 I carried out a multispecies ethnography involving participant observation of both human and non-human participants and semi-structured and informal interviews with a range of human actors within the cat fancy. All participants were members of the community or interested parties, either breeders, exhibiters, veterinarians and feline behaviourists, or involved with institution or show management.

The majority of participants were female which is representative of the cat fancy in general, as well as many other human-animal leisure activities, especially those involving dogs and horses (e.g. Baldwin & Norris, 1999; Dashper, 2017; Gillespie et al., 2002; Hultsman, 2012, 2015). A lack of ethnic diversity also appears to be representative of the cat fancy community. Age ranges of participants varied from the 20s to 70s, the majority being within the 40s to 60s. I was unable to secure contact with the over 80s who represent a small part of the fancy. This is possibly due to my recruitment methods being social media and the use of snowball sampling. I was careful to avoid the pitfalls of snowball sampling in relation to selecting an unrepresentative sample. For example, there is the risk of better known people being put forward or it can exclude those with smaller networks, such as introverts or those new to the cat fancy (Bernard, 2006). As such, I used additional selection techniques, including direct contact with members of online groups, posting requests for participation on online groups, and face-to-face convenience sampling at shows.

I conducted semi-structured interviews with 58 participants either by telephone, email, online messaging services such as Facebook Messenger, or in person depending on the logistics of distance and interviewee preference. I selected semi-structured interviews with the intent to enable 'rich personal narratives' to emerge (Sanders, 1999, p. xiii). My overall aim was to obtain in-depth narratives of people's lives within the cat fancy and their relationships with their cats.

During immersive participant observation at shows, I also spoke in person with countless breeders, exhibiters, and judges. In the interviews I allowed the participant to lead the conversation, following the rule, 'get people on to a topic of interest and get out of the way. Let the informant provide information that he or she thinks is important' (Bernard, 2006, p. 216). However, I also recognise that my participants and myself were co-authors of the interview (Montgomery, 2012).

I utilised a multi-site approach, following my participants to different locations within the UK for cat show events. Cat shows are public spaces, meaning that gatekeepers to this setting were not necessary for gaining access. As I became further integrated, I took on roles both as a steward assisting with the judging and accompanying participants during the exhibiting of their cats. Attending shows enabled me to gain first-hand insights into the social world of the cat fancy. As Brown (2016) found in her research on horses, shows provide the opportunity to witness the politics within judging, as well as to hear the conversations between people when discussing matters of importance to their way of thinking. Additionally, shows were useful for spending time with the cats and observing their behaviours and interactions with the aim of engaging with what matters to them.

I have been involved with human-cat cultures throughout my life, having always lived with cats and previously working within cat welfare and directly caring for cats at a rescue shelter. Through these experiences I have developed a knowledge of cat behaviour that has enabled me to recognise feline body language and to feel comfortable interacting with them (see Hurn, 2018). In other ways, the cat fancy was an unfamiliar context, a world of argot and social structures that I needed to learn in order to access the lived experiences of those involved. Overall my shared feline interest was a useful way of connecting with participants, and many would ask if I had cats of my own. It appeared important to them to know if I was also a 'cat person' and I could sense the relief when I would tell them about the cats with whom I share my life.

Consequences of engagement

Cat shows are held almost every weekend across the UK, usually at sports halls and community centres on the periphery of major towns. The halls are often large, open spaces, too cold in winter and too warm in summer, and bathed in bright artificial light. On entry to the show, one visually notices row upon row of metal cages, known as pens, temporarily housing the cats for the day. Around the edge of the hall are numerous tables selling cat-related wares and clothing, and cat products such as litter, food, and toys. People amble around amongst the rows of pens, conversing and gazing at the cats. Human voices dominate, with the occasional feline vocalisation and a background hum of ventilation systems. The smell is non-descript unless you wander by a rubbish bin and experience the waft of used litter. At the end of the hall, there is a row of decorated pens with large rosettes waiting for the best in show judging ceremony at the end of the day. Competitions organised under the Governing Council of the Cat Fancy (GCCF) run for one day, most often on Saturdays, while The International Cat Association (TICA), Cat Fancier's Association (CFA) and Fédération Internationale Féline (FIFe) shows typically run across the whole weekend.

Those involved in dog sports (Gillespie et al., 2002; Hultsman, 2012) and horse-related leisure (Dashper, 2017) are often seen by non-participants as 'crazy' due to the level of commitment and passion they devote to their hobby. To outsiders, these can be perceived as deviant activities (Gillespie et al., 2002) and participants can sometimes be seen as pathological (Dashper, 2017). Cat fanciers regularly describe themselves as 'crazy cat people' or feel that others perceive them as such. Antonia[2] explained that 'my family are not cat people. They all think I'm absolutely crazy', and Barbara said, 'everybody always knows me as the mad cat woman here, the neighbours think it's hilarious when they see me walking around the estate with the cats on the lead.' Meanwhile, others referred to themselves directly as 'crazy'. Kim expressed, 'it's lovely being in a show hall and ... for a change not being the crazy person who has a dozen cats'. Jessica asked me, 'Am I any different to any of the other breeders? Are we really all as crazy?'. In Geertz's (2005) famous ethnographic text, Balinese cockfighters also referred to themselves as 'cock crazy' in reference to the amount of time, resources, and commitment given to their cocks and cockfighting. In the cat world, this has largely become a positive marker of identity, as demonstrated by the almost five million Instagram posts with the hashtag #crazycatlady.

As people become immersed within their respective activity, they learn the subcultural and regulatory norms and rules within the social world and become embroiled into the 'community of practice' (Grasseni, 2005; Stebbins, 2015). Substantial investment is often required from participants particularly in terms of time, money, and personal relationships (Lamont et al., 2014). Monetary expenses include cat care products and services, travel to shows, nights at cat-friendly hotels, and the cost of entering the cat(s) into shows. For some, the associated financial investment is considered as prohibitive for regular showing. Olivia, for example, told me, 'I have only showed them four times so far and would have done more had it not been for the costs involved ... it's often more for one show taking four cats than it is for a two-week holiday to Florida'. Competitors may purchase equipment to enable participation and integration into the social group (Syrjälä, 2016). This acquisition is not enough in itself to represent the commitment required for becoming 'serious'. Instead it is the 'consumption, as well as on-going investment' that represents incorporation into the social world (Scammon, 1987, n.p). Individuals frequently make substantial changes to their everyday lives to accommodate their hobby. For example, with dog sports, participants often make life decisions based on the needs of both the dogs and the sport, including selecting cars that will be suitable for transporting the dogs and equipment to different events (Gillespie et al., 2002; Hultsman, 2012). This is also a factor for the cat fancy and especially for breeders. Becky explained that her 'house is pretty much already given over to the cats, so we've got floor to ceiling multi-level scratching posts, and all sorts going on.' This idea of the house being designed around the cats was also expressed by others, including Denise, who described the substantial changes that had been made to her home:

> It's sort of rather cat dominated ... The cats ... can go where they like. We have lots of scratching posts, climbing things for them all through the house and then in the garden ... we've got cat houses for them to sleep and eat, and keep warm in if they want to.

In addition to physical modifications, it is not unheard of for people to change or quit their occupations in order to accommodate their hobby (Hultsman, 2012, 2015). Antonia told me:

> I'm very lucky in that because I'm a stay at home Mum, I can be at home 24/7 for the kittens. A lot of breeders will book the week off work that their queen [female breeding cat] is due, so they'll lose out on annual leave to go on holiday to make sure that they're there for when the kittens are born.

This was confirmed by Alyssa: 'it's very time consuming ... In fact a friend of mine, she actually gave her job up in a school because the headteacher wouldn't let her have a week or two off to feed these kittens.' This demonstrates the sacrifices that participants may make for continuation of their involvement and ongoing cat care requirements. The resource constraints from developing a leisure career such as within the cat fancy are not a possibility for all, and this highlights the 'inequitable access to serious leisure' that can often exclude the disadvantaged (Gallant et al., 2013, p. 100). As such, involvement within the cat fancy can be seen as an indirect display of resources and capital.

The sociality of cat shows: company and competition

Cat exhibiting provides opportunities for progression within the social ranks and typically as people spend longer within the activity they take on additional roles. This can result in an escalation of involvement regardless of available resources, which in turn can cause conflict in external social relations, such as with family and friends (Stebbins, 2015). Individual leisure, for example, has been associated with a negative impact on marital relationships (Hodge et al., 2015). A study of dog sports found that conflict could arise when a participant attempted to balance family commitments with the hobby (Gillespie et al., 2002). Hultsman (2012) looked at couple involvement within dog agility but established that even where one member of the couple has attended shows to support and help build their relationship, the 'most meaningful interaction' is nonetheless commonly between the main competitor and their dog (p. 23). The time spent with their cats was hardly mentioned by the participants in this study. Unlike the dog world, my research suggests that motivations for participation in cat shows are not based on building relationships with the cats. The relations participants are seeking to create in the show hall are not ones of intersubjective fulfilment, rather, the cats facilitate access to resources and experiences for the cat fancier. Other human participants, not more-than-human ones, were perceived as being the main source of positive interactions.

Charles and Davies (2008) refer to animal shows as a 'means of establishing broader social networks' within a national network of others involved with breeding and showing (p.12). This was also a theme within cat showing, increasingly on an international scale with relationships developing across national boundaries with the importation and exportation of cats, and through cat fancy-related international online groups. Bradshaw (2017) argues that breeders and exhibiters see their cats and dogs as a means for meeting 'like-minded' people. This was also found by Hultsman (2012) with dog agility participants referring to their social interactions and friendships as reasons for their continued engagement. Many of my participants also described this as a driver behind their involvement. For example, Charlotte explained, 'it's quite nice because there are sort of lots of like-minded people there. You can talk cats and nobody minds'. Audrey told me: 'what I do like about the cat fancy is you meet all these like-minded people, but you don't know their backgrounds. You don't know who they are, and I think it's a great leveller'. Horses are seen as acting as similar conduits across typical social stratifications, such as age and status (Dashper, 2017). It could be suggested that rather than like-minded, cat fanciers are 'like-animaled', sharing similar orientations towards the same species that bring them together.

Despite the perceived positive social impacts from being a human member of the cat fancy, there are also negative interactions including interpersonal conflict. The level of competition within showing can be high and there are often fears of bias within judging (Baldwin & Norris, 1999; Gillespie et al., 2002). Success can bring some indirect financial reward through the promotion of breeding lines, but more significantly, it also brings status and recognition. According to Fox (2009, p. 98) winning animals are seen as a reflection of the breeder's skill in reproducing the breed standard. This connection between status and success is perpetuated through the registry naming system that results in breeders maintaining a named relation to winning cats even after the cat has been sold (see Hurn, 2008). Baldwin and Norris (1999) noted that winning competitions and the associated impact upon the validation of identity was considered one of the benefits of active membership with the American Kennel Club. As was also found by Fox (2009), despite most participants referring to enjoyment rather than winning as their motivation, many of those I met were disappointed not to win. This competitiveness was blamed for creating conflict and Dashper (2017) refers to how the 'pressures of competition exacerbate any divisions' that exist (p. 114). Denise told me, 'some people are very, very competitive. And they want their cats to win at all costs.' Sabrina noted that this competitiveness is also driven by people's emotional connection to their cats based on kinship: 'you're criticising someone's baby. You're saying that their baby is not good enough and people can get really cross about that.' This was also commented on by Scammon (1987) in relation to continued participation within horse competitions where the animal is seen as a marker of identity or extension of the self: 'self-identity now encompasses the horse. A compliment on how beautiful the horse is enhances your ego. A slur on her performance is a personal insult' (n.p, see Belk, 1996). Showing, then, is a nuanced activity with complex actual and potential benefits for participants.

However, this form of serious leisure can also entail costs at material, social and emotional levels. What, though, are the implications for the cat fancy's non-human participants?

What's in it for the cats?

At shows, the cats have to remain in their small metal pens, other than for judging, until the show has finished. Each cat is allowed a white litter tray, white blanket or bed, and white water bowl. Each item must be white in an attempt to ensure anonymity. The public are given access to the show hall in the afternoon after the main judging has finished and during this period, visitors are able to 'gaze' upon each of the cats in their pens and talk with the exhibiters. Generally, the public spectate from a distance, with physical contact with the cats being minimised due to fears of contamination of infectious diseases between the cats. In this respect, the show acts as a form of visual consumption, with a commercial and voyeuristic angle acting as entertainment, similar to other animal-based leisure productions including aquariums, zoos, and rodeos (Beardsworth & Bryman, 2001; Milstein, 2009; Peñaloza, 2001). Building on Urry's (1990) concept of the 'tourist gaze', this visual consumption of nonhuman animals has been referred to as the 'zoological gaze' (Franklin, 1999). The pen, set up with no hiding spaces, presents minimal opportunities for the cats to avoid the human gaze, in the process demonstrating human power over the powerless (Milstein, 2009, p. 32). This anthropocentrism is further demonstrated by a preference of some judges against covered beds which can help reduce stress (see Gourkow, 2016; Trevorrow, 2013) but make it more difficult to remove the cat from the pen for judging.

During the judging, cats are taken from their pens on multiple occasions to be evaluated by different people. They are held in place on a trolley by a steward for the judge to make their assessment on aesthetic conformation, and depending on the breed of the cat or show class, also on temperament. The judge will run their hands over the cat, look at his or her coat, lengthen the tail, as well as check key features for the required breed characteristics known as 'the standard of points'. This handling can be stressful for some individuals. This was described by Gail, a cat behaviourist, who told me about her concerns with shows:

> It's the taking out, the handling, the touching, the holding, the stretching … a lot of judges sort of complained of being nailed by the cats and I'm thinking I'm not blooming well surprised. I would expect to be harmed in some way if I was to use such overtly disrespectful handling.

On one occasion during my fieldwork, I heard a cat hiss and yowl at the top of her voice and the clatter of a quickly closing metal cage door. I did not see the actual incident, but the steward had received a scratch from the cat during the process. Later in the day, I went to see the cat in question. According to her caregiver, she was distressed for the rest of the show and she also had a mark on her face, possibly from being put back into the cage at speed. Not all interactions are negative, however. Judges also play with the cats using feather toys and some will speak to them and ask them questions, such as, 'aren't you lovely?' and 'you're three months old, right?' According to Irvine and Cilia (2017) these processes of speaking with and for animals 'provide evidence that people see pets as subjective beings with interests and reciprocal roles in the interaction' (p. 4). This has also been recorded in veterinary clinics with guardians speaking on behalf of their companion animals and their symptoms (Arluke & Sanders, 1996; Sanders, 2003). Throughout my fieldwork, judges commonly answered the cats vocalised complaints by saying, 'I know'. A similar response has been noted for vets reacting to distress calls during examinations as a way of indicating their understanding of the animal's experience and building relationships with the animal's human guardians (MacMartin, Coe, & Adams, 2014).

Despite the focus on conformation, judges are sometimes influenced by emotion through interaction with the cats, for example a cat which reacts positively and demonstrates affiliative behaviours may influence the judge towards making a more positive evaluation, while a cat which demonstrates agency through resistance in the form of biting or scratching may be disqualified.

There are multiple factors influencing judging decisions and although human-dominated, it is not simply a one-way process with the role of intersubjectivity likely to prevent the cats being seen purely as 'ornaments' for aesthetic pleasure. However, although this feline agency can have some influence over judging decisions, this does not result in further attempts within the cat fancy to enable the conditions for greater expression of feline agency and show environments can be argued to limit the cat's behavioural repertoire.

Wandering between the pens, one notices the array of physical and emotional reactions from the cats to being in the show environment. Some are visibly stressed with ears flat and huddled in a corner, others are fighting for attention from the public by rolling on their backs, meowing, or pawing at the bars and seeking escape, while still others are hidden under their blankets and many sitting motionless, displaying possible signs of inhibition or learned helplessness (see Carney & Gourkow, 2016; Sundahl, Rodan, & Heath, 2016). As one veterinarian, Mia, explained to me,

> Cat behaviour is really difficult to understand and it's really misunderstood by a lot of cat lovers. And because they withdraw and freeze sometimes when they're in situations that are out of their control, I think there's probably thousands of cat breeders and showers who have got cats who are bloody terrified and they think that they're just sitting there chilled.

The umwelt of an organism is influenced by its physical characteristics, so that different species have distinctive experiences of the same environment (Von Uexküll, 2010; Warkentin, 2009). Due to the feline umwelt and sensory abilities, the show environment will be particularly challenging to many cats. Owing to their territorial nature, cats place importance upon familiar localities, making travel and new environments stress-inducing (Atkinson, 2018). This is exacerbated as part of the wider show experience, as human and feline participants regularly travel hundreds of miles, even internationally, to attend shows with the cats confined to travel crates. Additionally, competitors often stay overnight in hotels, with any cats who are likely to spray restricted to the bathrooms to prevent incurring cleaning charges. At the show, attempts are made to prevent visual interaction between cats located in pens next to one another with the pens decorated with drapes on the sides. However, this intervention is based on an ocularcentrism and cats are much more reliant than humans on the olfactory system for investigating novel objects and communicating with others (Bradshaw, Casey, & Brown, 2012). They also have a sensitive hearing for prey detection purposes which may make certain noises more noticeable (Atkinson, 2018). At a non-GCCF show, I visited, judging pens were set up in an 'L' shape. This meant the cats in the pens located at the corner of the 'L' were facing each other. As the cats were being put into the pens by their guardians, there was an extremely loud vocalisation of hissing and yowling made by a male cat who was clearly upset about being so physically close to another cat. I later spoke to the cat's guardian, who explained, partly in jest, that he 'didn't like longhaired cats'. In general, environmental enrichment and opportunities for cats to express natural behaviours at shows are minimal despite evidence linking environmental complexity, such as hiding spaces and vertical places with an improvement in feline well-being and stress reduction (Gourkow, 2016; Kry & Casey, 2007; Vinke, Godijn, & van der Leij, 2014). As such, there is a failure within showing practices to attempt to engage with the cat's way of being.

Despite this, many exhibiters spend time interacting with their cats and holding them beside their pens and the bond between some of them is evident. The cats may meow and come to the front of their pen to greet the familiar face, while the human may speak gently to the cat and congratulate them on their win from the morning's judging. The cat is often rewarded with some attention, some special food, or a new toy purchased from one of the show vendors before the human returns to his or her socialisations with other human competitors. However, some choose to set up foldable chairs in front of their pen and spend the afternoon with their cats, often accompanied by grooming and play using feather toys. This is likely in part due to a fear of sabotage and cruelty towards their cats by jealous competitors. Pamela conveyed stories of the harm that she had seen or heard being done to cats, 'poisoning is the most common but

I remember a cat's ear tip being cut off, dye squirted on the coat, chunks of fur snipped off ... Kittens have been stolen from shows. It's rare but it happens.' However, I believe for some people they also stay with their cats because they want to spend time with them. Alyssa told me that, 'I like to spend my time at the show with my cat because I don't want him to be sat there on his own'. Furthermore, Hayley explained, 'I personally like to be with my cats all day ... the point is I'm showing my cats, and if I was showing dogs I'd be with them all day'. Nevertheless, these comments were in the minority, and in contrast to what has been found for dog sports (e.g. Hultsman, 2012), the opportunity for exhibiters to bond with their cats or a reference to the cat's enjoyment has generally not featured as a motivator for partaking. Only Kylie referred to her cat's enjoyment as justification for the activity:

> I would not take my cat unless he enjoyed it. You can tell if a cat enjoys it ... mine actually loves a day out ... he is so relaxed and laidback at the show ... The day that he doesn't want to get into his box or doesn't want to go into the car ... I know that's the day I stop showing him.

When asked why they show, the majority of my interviewees referred to their own motivations, rather than making reference to the cat's experience. Reasons given usually related to the human social interactions or learning more about what judges are looking for. For example, Jessica explained that she exhibits, 'to know that you're keeping up, that your cats still conform to the standard. And it's sociable. It's people. It's nice to see what other people are breeding.' This was supported by Yvonne, who said, 'it's fun and you get ... to compare your cat with other people's cats. See what else is out there ... And you also meet contacts and meet other people'.

Some of the participants in my research did refer to lack of enjoyment expressed by cats brought to shows as a point of critique towards fellow competitors, often with accusations of them being over-competitive. Kylie described such an experience:

> I had one next to me the other day that was panting away, he had an upset stomach and they said, 'oh he doesn't like travelling in the car', I said, 'why did you bring it?' 'Oh, we need one more to get an imperial [certificate], well why are you doing it? It's not fair on the cat.

This opinion was also shared by Becky:

> I think there's people who should look more to the welfare of the cats than their own need for a rosette. And I see that as a steward ... as I'm seeing the cats when people are out of the hall ... some are just plain unhappy and shouldn't be there.

As Haraway (2008a) found with those using aversive training techniques for agility dogs, people will be the 'subject of disapproving gossip' if their norms of care do not correspond with the wider group (p.447). Many participants informed me that as soon as their cats stop demonstrating behaviours that in their opinion indicated enjoyment that they would be retired from showing. Denise was one example:

> We have several cats that are very good types, have done very well at shows ... and then have been taken to another show for the next level and they're not happy. So we stop showing them because they obviously don't want to be shown anymore ... I want the cat to be happy. It's not all about getting titles and winning prizes for us.

The motivation for retiring cats is not only driven by welfare and individual-level concerns though. It can also be motivated by chance of success, as indicated by Victoria:

> I have had cats who haven't particularly enjoyed it, but you just retire them because there's no point taking an unhappy cat ... because part of the judging is the temperament, so you're looking for a happy, handleable animal, not something that wishes it wasn't there.

In the above comment, Victoria also demonstrates the role of feline agency in creating the conditions for winning. Sophia told me about her cat, Pepper, who demonstrated agency through preference and resistance that was likely also influenced by fear and stress:

> Pepper ... was an excellent show cat. He sailed through showing ... until he met a judge that he did not like and from then on he was fine ... with anyone else but not this judge. He would always tense up at this judge, even just being near him. Eventually he started to worry about showing and started to get growly and upset on the show bench.

Displays of resistance have often been used to describe animal agency within human-animal 'contact zones' (Haraway, 2008b e.g. Warkentin, 2009). Sophia referred to another of her cat's responses to showing in a way that indicated a perception of agency through resistance: 'Poppy decided showing was not for her ... She showed us this when she bit a judge and that told us that this was not the thing that she wanted to do in life' so they stopped showing her. However, the success of resistance is also bound up within wider power relations and the actions of humans (Carter & Charles, 2013), as demonstrated by Antonia:

> At home he is the most placid cat, he will cuddle up next to me and he'll be on his back with his legs up in the air, demanding belly rubs ... Soon as he goes to a show, he's hissing at me, he won't come out, he hates coming out of the pen. But that's because he's in his teenage phase at the moment.

The actions taken by Antonia's cat could be seen as an expression of agency and resistance. Conversely, Antonia believes this is due to him experiencing a 'teenage phase' so she is choosing to ignore his agency and continuing to show him. This demonstrates how non-human agency can be limited by human power (Dashper, 2017). It is clear that some cats are still presented at shows despite recognition by participants that the show environment is not suitable for all cats and regardless of signs of excessive stress and lack of enjoyment, in the process prioritising human desires to socialise, compete and win prizes.

Conclusion

In this article I have extended the concept of 'serious leisure' through the lens of the cat fancy, critiquing an overly positive and anthropocentric approach that neglects the associated costs incurred by those involved, especially the other-than-human animals (see Gallant et al., 2013). For many participants, the cat fancy and cat shows provide social interaction, opportunities to develop knowledge and skills and to construct identities. Yet, there are also substantial consequences of engagement as a result of the levels of commitment and resources required for 'career' progression, in addition to interpersonal conflict and experiences with competitive behaviour. I would argue that often in the cat fancy, human leisure needs are prioritised over the well-being of the feline individuals with the cats becoming embroiled in human conflict. Due to the feline umwelt, exhibition spaces and show requirements present considerable challenges to many cats. This was recognised by some of my participants, but often the needs of the cats only arose as a point of criticism towards other human exhibiters.

Cats and the idea of cats constitute a substantial part of individual and group subjectivity, with participants in the cat fancy sharing a 'like-animaled' connection. However, the social aspects of cat showing are largely anthropocentric and focused on human–human interactions, with the cats acting as conduits for these relations and as markers of identity. This article has attempted to answer Birke's (2009) question, 'what's in it for the animals?' (p. 1) by interrogating the concept of serious leisure in relation to its implications for feline participants. It could be suggested that show cats lead a relatively pampered life compared to many others, receiving human attention, special food and toys, as well as cat-orientated alterations in the home. However, as I have demonstrated, the practices that enable the cat showing activity to take place can be detrimental to cats and limit, or even deny altogether, the expression of agency. This is not to state that there are not relations based on human-cat intersubjectivity, but the focus of the activity and its power dynamics results in human interests being prioritised.

Notes

1. Research was conducted at a range of organisations running shows in the UK. However, as the predominant organisation in the UK, most of the fieldwork relates to the Governing Council of the Cat Fancy's (GCCF) style of shows.
2. Pseudonyms are used for both human and more-than-human participants.

Disclosure statement

No potential conflict of interest was reported by the author.

ORCID

Emily Stone ⓘD http://orcid.org/0000-0002-7409-982X

References

Arluke, A., & Sanders, C. (1996). *Regarding animals*. Philadelphia: Temple University Press.
Atkinson, T. (2018). *Practical feline behaviour: Understanding cat behaviour and improving welfare*. Wallingford: CABI.
Baldwin, C. K., & Norris, P. A. (1999). Exploring the dimensions of serious leisure: 'Love Me – Love My Dog!'. *Journal of Leisure Research*, *31*(1), 1–17.
Beardsworth, A., & Bryman, A. (2001). The wild animal in late modernity. *Tourist Studies*, *1*(1), 83–104.
Belk, R. W. (1996). Metaphoric relationships with pets. *Society and Animals*, *4*(2), 121–145.
Bernard, H. R. (2006). *Research methods in anthropology: Qualitative and quantitative approaches*. Lanham: Altamira Press.
Birke, L. (2009). Naming Names – Or, What's in it for the Animals? *Humanimalia*, *1*(1), 1–9.
Bradshaw, J. W. S. (2017). *The animals among us: The new science of anthrozoology*. London: Penguin Books.
Bradshaw, J. W. S., Casey, R. A., & Brown, S. L. (2012). *The behaviour of the domestic cat*. 2nd edn. Wallingford: CABI.
Brown, C. (2016). From working to winning: The shifting symbolic value of connemara ponies in the west of Ireland. In D. L. Davis & A. Maurstad (Eds.), *The meaning of horses: Biosocial encounters* (pp. 69-84). London: Routledge.
Carney, H., & Gourkow, N. (2016). Impact of stress and distress on cat behaviour and body language. In S. Ellis & A. Sparkes (Eds.), *Feline stress and health: Managing negative emotions to improve feline health and wellbeing* (pp. 31-39). Tisbury: International Cat Care.
Carr, N. (2014). *Dogs in the leisure experience*. Wallingford: CABI.
Carr, N. (ed). (2015). *Domestic animals and leisure*. Basingstoke, UK: Palgrave Macmillan.
Carter, B., & Charles, N. (2013). Animals, agency and resistance. *Journal for the Theory of Social Behaviour*, *43*(3), 322–340.
Charles, N., & Davies, C. A. (2008). My family and other animals: Pets as kin. *Sociological Research Online*, *13*(5), 4.
Dashper, K. (2017). *Human-animal relationships in equestrian sport and leisure*. Abingdon: Routledge.
Fox, R. (2009). Pedigree and breeding: Love, status and control. In T. T. Holmberg (Ed.), *Investigating human/animal relations in science, culture and work* (pp. 97-113). Uppsala: Uppsala universitet.
Franklin, A. (1999). *Animals & modern cultures: A sociology of human-animal relations in modernity*. London: Sage Publications.
Gallant, K., Arai, S., & Smale, B. (2013). Celebrating, challenging and re-envisioning serious leisure. *Leisure/Loisir*, *37*(2), 91–109.
Geertz, C. (2005). Deep play: Notes on the Balinese cockfight. *Daedalus*, *134*(4), 56–86.
Gillespie, D. L., Leffler, A., & Lerner, E. (2002). If it Weren't for my Hobby, I'd have a life: Dog sports, serious leisure, and boundary negotiations. *Leisure Studies*, *21*(3–4), 285–304.

Gourkow, N. (2016). Prevention and management of stress and distress for cats in homing centres. In S. Ellis & A. Sparkes (Eds.), *Feline stress and health: Managing negative emotions to improve feline health and wellbeing* (pp. 89-102). Tisbury: International Cat Care.

Grasseni, C. (2005). Designer cows: The practice of cattle breeding between skill and standardisation. *Society & Animals*, 13(1), 33–49.

Green, B. C. (2001). Leveraging subculture and identity to promote sport events. *Sport Management Review*, 4(1), 1–19.

Gruen, L. (2015). *Entangled empathy: An alternative ethic for our relationships with animals*. New York: Lantern Books.

Haraway, D. (2008a). Training in the contact zone: Power, play, and invention in the sport of agility. In B. Da Costa & K. Philip (Eds.), *Tactical biopolitics: Art, activism, and technoscience* (pp. 445-464). Cambridge, MA: MIT press.

Haraway, D. (2008b). *When species meet*. Minneapolis and London: University of Minnesota Press.

Hirschman, E. C. (1994). Consumers and their animal companions. *Journal of Consumer Research*, 20(4), 616–632.

Hodge, C., Bocarro, J. N., Henderson, K. A., Zabriskie, R., Parcel, T. L., & Kanters, M. A. (2015). Family leisure: An integrative review of research from select journals. *Journal of Leisure Research*, 47(5), 577–600.

Hultsman, W. (2015). Dogs and companion/performance sport: Unique social worlds, serious leisure enthusiasts, and solid human-canine partnerships. In N. Carr (Ed.), *Domestic animals and leisure* (pp. 35-66). London: Palgrave Macmillan.

Hultsman, W. Z. (2012). Couple involvement in serious leisure: examining participation in dog agility. *Leisure Studies*, 31(2), 231–253.

Hurn, S. (2008). What's love got to do with it? The interplay of sex and gender in the commercial breeding of welsh cobs. *Society and Animals*, 16(1), 23–44.

Hurn, S. (2018). Encounters with dogs as an exercise in analysing multi-species ethnography. In J. Lewis (Ed.), *SAGE research methods dataset*. London: SAGE. doi:10.4135/9781526440921

Irvine, L., & Cilia, L. (2017). More-than-human families: Pets, people, and practices in multispecies households. *Sociology Compass*, 11(2), e12455.

Kry, K., & Casey, R. (2007). The effect of hiding enrichment on stress levels and behaviour of domestic cats (*Felis sylvestris catus*) in a shelter setting and the implications for adoption potential. *Animal Welfare*, 16(3), 375–383.

Lamont, M., Kennelly, M., & Moyle, B. (2014). Costs and perseverance in serious leisure careers. *Leisure Sciences*, 36(2), 144–160.

Lloro-Bidart, T. (2018). A feminist posthumanist multispecies ethnography for educational studies. *Educational Studies*, 54(3), 253–270.

MacMartin, C., Coe, J. B., & Adams, C. L. (2014). Treating distressed animals as participants: *I know* responses in veterinarians' pet-directed talk. *Research on Language and Social Interaction*, 47(2), 151–174.

Milstein, T. (2009). 'Somethin' Tells Me It's All Happening at the Zoo': Discourse, power, and conservationism. *Environmental Communication: A Journal of Nature and Culture*, 3(1), 25–48.

Milton, K. (2005). Anthropomorphism or Egomorphism? The perception of non human persons by human ones. In J. Knight (Ed.), *Animals in person: Cultural perspectives on human-animal intimacies* (pp. 255-271). Berg: Oxford.

Montgomery, A. (2012). Difficult moments in the ethnographic interview: Vulnerability, silence and rapport. In J. Skinner. (Ed.), *The interview: An ethnographic approach* (pp. 143-159). ASA Monographs 49. London: Bloomsbury.

Peñaloza, L. (2001). Consuming the American West: Animating cultural meaning and memory at a stock show and rodeo. *Journal of Consumer Research*, 28(3), 369–398.

Rojek, C. (2001). Preface. In R. Stebbins (Ed.), *New directions in the theory and research of serious leisure* (pp. i-iv). Lewiston, NY: Edward Mellen Press.

Sanders, C. (1999). *Understanding Dogs: Living and working with canine companions*. Philadelphia: Temple University Press.

Sanders, C. (2003). Actions speak louder than words: Close relationships between humans and Nonhuman Animals. *Symbolic Interaction*, 26(3), 405–426.

Scammon, D. L. (1987). Breeding, training, and riding: The serious side of horsing around. *Advances in Consumer Research*, 14, 125-128.

Stebbins, R. A. (1982). Serious leisure: A conceptual statement. *The Pacific Sociological Review*, 25(2), 251–272.

Stebbins, R. A. (2015). *Serious leisure: A perspective for our time*. New Brunswick: Transaction Publishers.

Sundahl, E., Rodan, I., & Heath, S. (2016). Providing feline-friendly consultations. In I. Rodan & S. Heath (Eds.), *Feline behavioural health and welfare* (pp. 269-286). St Louis, MO: Elsevier.

Syrjälä, H. (2016). Turning point of transformation: Consumer communities, identity projects and becoming a serious Dog Hobbyist. *Journal of Business Research*, 69(1), 177–190.

Trevorrow, N. (2013). Helping cats cope with stress in veterinary practice. *Veterinary Nursing Journal*, 28(10), 327–329.

Urry, J. (1990). *The Tourist Gaze: Leisure and travel in contemporary societies*. London: SAGE.

Vinke, C. M., Godijn, L. M., & van der Leij, W. J. R. (2014). Will a hiding box provide stress reduction for shelter cats? *Applied Animal Behaviour Science, 160,* 86–93.

Von Uexküll, J. (2010). *A foray into the worlds of animals and humans.* Minneapolis: Minnesota University Press.

Warkentin, T. (2009). Whale Agency: Affordances and acts of resistance in captive environments. In S. E. McFarland & R. Hediger (Eds.), *Animals and agency: An interdisciplinary exploration* (pp. 23-43). Leiden: Brill.

Warkentin, T. (2010). Interspecies Etiquette: An ethics of paying attention to animals. *Ethics and the Environment, 15*(1), 101–121.

An ecological-phenomenological perspective on multispecies leisure and the horse-human relationship in events

Katherine Dashper (iD) and Eric Brymer (iD)

ABSTRACT

More-than-human approaches open up theoretical and methodological space for considering if and how all animals, human and nonhuman, play important roles in shaping relationships, actions and encounters in leisure. This paper introduces an ecological-phenomenological framework for understanding relationships between animate actors and their environment in and through leisure. The example of human riders and horses in the context of a pleasure ride leisure event is used to illustrate the application of the framework for understanding the importance of individual differences and constraints, and their interaction with the environment, in appreciating the variety of affordances and possible outcomes in leisure practices. The ecological-phenomenological framework has theoretical and methodological implications for researchers of multispecies leisure, and may have practical application for event managers and designers of multispecies leisure activities. This article is important because it transforms current appreciation of multispecies leisure and opens doors to new ways of thinking and investigating the value and meaning of leisure in a multispecies context.

Introduction

Leisure studies is a field dominated by anthropocentric, humanist approaches which prioritise human perspectives and actions, and downplay nonhuman experiences (Dashper, 2018). However, the 'animal turn' that is affecting many other social science disciplines and subject areas is starting to be recognised in leisure studies as well. Leisure cannot be seen as a wholly human practice, and empirical studies and theoretical analyses are beginning to recognise the varied and often significant roles played by animate nonhumans in leisure spaces and practices. A variety of approaches have been adopted for considering if and how leisure can be understood as a more-than-human phenomenon involving animate nonhumans, in order to open up space for explorations of some of the difficult, messy and complicated interactions that constitute leisure (and other practices) in multispecies worlds.

There are a variety of theoretical and methodological perspectives that draw on more-than-human perspectives that can be used to explore multispecies and more-than-human leisure, challenging the dominance of humanism in the social sciences and seeking to decentre human experiences and human actions as the main foci of research (Badmington, 2003). These theories 'challenge human exceptionalism, posit that human-nonhuman relations/relationships emerge temporally, and/or demonstrate how what we ontologically understand as "human" is really a complex relation with other species' (Lloro-Bidart, 2017, p. 113). Such perspectives open up space for considering if and how nonhumans (animate and inanimate) play key roles in different spaces, contexts and interactions, and for exploring some of the complex interplay between humans and nonhumans in different settings. In so doing, researchers operating within more-than-human frameworks are required to challenge traditional ways of thinking about and doing research, in order to try and incorporate nonhuman others as actors and subjects, a project which 'entails challenging, and moving away from, the privileging of the speaking, rationally reflective human agent/research that continues, implicitly at least, to frame knowledge production in the social sciences and humanities' (Dowling, Lloyd, & Suchet-Pearson, 2017, p. 827).

In this paper we introduce an ecological-phenomenological approach as a fruitful tool for understanding multispecies leisure as a dynamic relationship between animate perceivers (human and nonhuman) and environment. This relational approach may provide a useful framework for understanding some of the complexities of leisure involving multiple species in a variety of environments. We use the example of horse-riding leisure events – 'pleasure rides' – to illustrate some of the theoretical and practical applications of this framework in the context of multispecies leisure.

Current approaches in multispecies leisure research

Two of the dominant conceptual frameworks for researching multispecies leisure are Actor Network Theory (ANT) and human-animal studies. Both draw on more-than-human insights to consider the important roles that nonhumans play in leisure, but do so in different ways and with different emphases. Both ANT and human-animal studies offer interesting and informative frameworks for researching leisure as a multispecies practice, but both also have limitations for considering human-nonhuman interactions, some of which are addressed by the ecological-phenomenological approach we introduce below.

Actor network theory (ANT) has been a popular perspective adopted in leisure and tourism studies, and represents one useful way for approaching multispecies leisure. Described by its proponents as less of a theory and more of a method or a 'toolbox', ANT focuses less on the 'why' questions of social science, and more on the 'how' – 'how it [tourism in this case] is assembled, enacted, and ordered; how it holds together; and how it may fall apart' (van der Duim, Ren, & Jóhannesson, 2013, p. 5). One of the basic premises of ANT is the principle of general symmetry: analytically, all actors – human and nonhuman – are supposed to be treated in the same way, and are seen as equally able to create effects (van der Duim, Ren, & Jóhannesson, 2017). The methodological result of this approach is that no assumptions can be made in advance about who or what will act in any given circumstance (Van der Duim, 2007). The researcher is encouraged to 'follow the actor' to understand how networks are brought into being, how they develop and how they may disintegrate. The researcher cannot know in advance which actors in leisure networks are of most significance, so the task is to describe relations in the network.

Applied to leisure studies this suggests that nonhumans can act and have effects on others (including humans) in leisure spaces and contexts, but that to understand what those effects may be will require careful empirical investigation and analysis. Ethnography is a commonly used approach in ANT-influenced studies, as researchers can remain open to possibilities of different actors – human and nonhuman – acting in surprising and unexpected ways (Beard et al., 2016; Lamers, van der Duim, & Spaargaren, 2017). As Sayes (2014, p. 145) argues, ANT 'asks that we remain open to the possibility that nonhumans add something that is of sociological relevance to a chain of events: that something happens, that this something is added by a nonhuman, and that this addition falls under the general rubric of action and agency.'

ANT is thus a useful position from which to explore more-than-human aspects of tourism and leisure, but it has limitations in terms of understanding leisure as a multispecies phenomenon, where human and nonhuman *animals* act together, sometimes separately, and sometimes in opposition. One of the key features of ANT that make it a distinct and powerful approach from which to explore issues from a more-than-human position is perhaps also its biggest weakness in terms of trying to understand interspecies interactions. General symmetry is fundamental to decentring human experience and overcoming human exceptionalism, as it positions all actors – human and nonhuman, animate and inanimate – as analytically equal. However in doing so, individuals and specific interspecies encounters and relationships can disappear and nonhuman animals in particular risk fading from focus (Dashper, 2018). As Cohen and Cohen (2017, p. 9) argue, within many ANT studies 'live animals seem to lose their status as sentient beings, paradoxically within the very approaches that advocate a posthumanist ontology'.

Human-animal studies offers a different position from which to consider multispecies phenomena and begins to overcome this limitation in ANT. Still building on more-than-human assumptions and goals, human-animal studies focuses specifically on interactions and relationships between human and nonhuman *animals* and leaves other materials and objects to take a backseat in analysis. Human-animal studies is a broad field, and researchers within this area adopt a variety of theoretical positions and come from diverse disciplinary backgrounds in anthropology, sociology, geography and management, amongst others. However, what binds this approach together is focus on the interactions between human and nonhuman animals, and, as Haraway (2008, p. 66) puts it, 'the fleshy, historical reality of face-to-face, body-to-body subject making across species' that take place within different spaces and contexts. As with ANT, human-animal studies scholars often adopt ethnographic methods to study interspecies encounters, and the subpractice of multispecies ethnography is a multidisciplinary endeavour through which researchers are 'studying contact zones where lines separating nature from culture have broken down, where encounters between *homo sapiens* and other beings generate mutual ecologies and coproduced niches' (Kirksey & Helmreich, 2010, p. 546). Attempts are made to recognise nonhuman animals as actors in their encounters with humans and to try to understand nonhumans as far as possible on their own terms. Within leisure studies, Dashper (2017) has explored human-horse relationships and the deeply embodied, non-verbal interactions that take place as rider and horse try to negotiate complex tasks together. Charles (Charles, 2014; Charles & Davies, 2008) has studied pet keeping, whilst Gillespie, Leffler, and Lerner (2002) and Carr (2014) consider dog agility and other forms of human-dog leisure. These, and other studies, make nonhuman animals visible and central in the research, and focus on the lived, embodied experiences of multispecies leisure.

However, although multispecies ethnography and human-animal studies offer exciting avenues for researching multispecies leisure, they are not without their limitations. For all good intentions to include nonhuman animals as active actors in interspecies encounters, multispecies ethnography retains links to its humanist ethnographic home through focus on verbal and written language, on what can be seen and understood by a human researcher, and the necessity to communicate multispecies encounters in forms accepted and understood within the academy. This limits the extent to which the 'voices' of nonhuman animals can really be heard, understood and represented within multispecies ethnography (Dashper, 2017; Madden, 2014). Multispecies researchers, whilst openly committed to the more-than-human project to decentre human actions and account for nonhuman experiences, often struggle to actually do this in practice. Pacini-Ketchabaw, Taylor, and Blaise (2016) discuss some of these challenges in their own work, and call on scholars to continue to push boundaries, theoretically and methodologically, arguing that 'making the shift from representing animals as objects of study to engaging with animals as active research subjects requires a different set of habits, skills and dispositions' (p.156). This is extremely challenging, and researchers continue to experiment through drawing on ethology and autoethnography, amongst other tools and positions, to try to understand nonhuman animal experiences in more-than-human ways (Birke & Hockenhull, 2015; Dashper, 2017).

This brief consideration of two key approaches to researching multispecies encounters illustrates that this is a dynamic, evolving and challenging field, with researchers experimenting with method, approach and representation in an effort to overcome human exceptionalism and account for the importance of nonhuman actors. In the next section we introduce an ecological-phenomenological framework as another fruitful approach for trying to understand multispecies leisure and one that can begin to overcome some of the limitations in both ANT approaches and human-animal studies as briefly outlined here.

Introducing an ecological-phenomenological approach to multispecies leisure

Ecological psychology stemmed from the desire to understand human behaviour and cognition, and the realisation that the traditional approach that emphasised the individual and cognitive structures was limited (Gibson, 1979). Ecological psychology undermined the Realist-Cartesian paradigm and shifted the emphasis to recognise the importance of the animal-environment relationship in behaviour. As a consequence the processes underlying behaviour were considered consistent across all animal (including human) life. The ecological psychology approach is predicated on the animal (animate perceiver)-environment relationship. Rather than seeing cognition, behaviour and so forth as rooted in the cognitive functions of individual (generally human) actors, the ecological psychology approach recognises the role of the environment. The notion stems from the realisation that as animal and world evolved together and are interdependent it is important to consider them in relation to each other (Costall, 2001). The animal-environment mutuality as the scale of analysis for understanding multispecies leisure provides an opportunity to address the role of individual differences across all animal species (including human animals); rather than assuming a 'species' focus it allows for the capacity to think in terms of singular animals (Lestel, Bussolini, & Chrulew, 2014). Behaviour emerges from individual animals as they attempt to satisfy a range of individual, task and environmental constraints at any moment in time (Davids, Button, & Bennett, 2008). In recent years the ecological psychology approach has been expanded through the addition of phenomenological concepts and approaches to investigating experience (Immonen, Brymer, Davids, Liukkonen, & Jaakkola, 2018).

The complex relationships indicated by animal-environment systems requires a phenomenological approach to make visible otherwise hidden meaning (Withagen et al., 2017). Similar to the ecological approach in psychology, phenomenology does not follow the traditional positivist approach to understanding phenomenona (Brymer & Schweitzer, 2017). For the most part in modern times phenomenological analyses have dealt with human lived experience. However, as Martin and Peñaranda (2001) point out, humans are animals and the phenomenological method has been extended to investigating the *lifeworld* of nonhuman animals. According to Lestel et al. (2014) Husserl explicitly referred to all animate life when explicating his methodology. Phenomenology, considers consciousness as intentional, which means that consciousness and cognition are always towards something (Brymer & Schweitzer, 2017).

Intentionality is present in the lived worlds of many animals. Lestel et al. (2014) exemplified this process through describing the knot tying activities of an Orang-utan. The significance of this notion implies that all animal behaviour involves meaningful relations between the animal and its environment, as opposed to causal or mechanistic interactions. We share our world not just with other humans, but with other animals who are intentional, responsive, interpreting agents (Painter & Lotz, 2007). Aspects of the phenomenological method, such as bracketing or the setting aside the taken for granted scientific or naturalistic attitudes towards all other animate beings, opens up the possibility to intuit intentionality in the other as embodied agent. As researchers, phenomenology opens up ways of being and understanding the 'other' and a process that helps enhance our capacity to be sensitive to the 'distinctiveness of particular embodied souls and the intelligible intentionality and subjectivity they manifest' (Lestel et al., p.138). As Ruonakoski (p. 77) pointed out:

> just as we cannot live the experience of a poet of the 16th Century, neither can we capture the experience of a chimpanzee, a parrot or a gorilla, without any mediations. We can, however, without abandoning the standards of scientific rigor, give ourselves over to the task and the project of interpretation, and in so doing, we can be open to non-human animal others.

Both the phenomenological and ecological frameworks have critiqued the Realist-Cartesian paradigm in favour of a relational approach. While not yet applied to multispecies leisure, this approach has become established in human activities such as sport, learning and behaviour change (Brymer & Davids, 2013; Brymer, Davids, & Mallabon, 2014). In this section we show how this combined approach is ideally suited to the multispecies leisure field because it proposes that the interactive relationship between the individual animal and the environment is a relevant scale of analysis for understanding animal (including human) interactions. Key concepts from this approach include the notions of constraints and affordances.

Constraints

Constraints are boundaries which shape the emergence of behaviours (Newell, 1986). The interaction of different constraints guides the animal to seek stable and effective patterns of behaviour during goal-directed activity, which satisfies these constraints. Constraints have been classified as individual, environment and task.

Individual constraints are the unique structural and functional characteristics of each animal and include attributes related to their historical, physical, psychological, cognitive and emotional make up. In ecological psychology, the individual is conceptualised as exemplifying a complex, open system in nature. Such a system is defined as containing a number of interacting constituent parts or dimensions, all capable of interacting and influencing system behaviours over time. An animal's body shape, fitness level, age and psychological factors may shape the way the individual animal approaches a task. Individual constraints also include previous experiences, needs, interests, meanings and patterns of behaviour (Maiteny, 2002). These individual factors provide affordances (see next section) for action and play a significant and important role in determining the behaviours adopted by individual animals. Individuals are described as active agents with different individual constraints that illustrate the distinct strategies that may be used to solve problems. The solutions which emerge from the activities of different individual animals present important implications for the design of multispecies leisure experiences and for understanding multispecies leisure activities. These unique individual characteristics can be viewed as resources for each animal that channel the perception of information and the solving of particular task problems. These relatively unique characteristics can lead to individual-specific adaptations. An individual's functional behaviours in response to task challenges aim to satisfy his/her own unique constraints. Variability in behaviour is to be expected and can play a functional role as each individual seeks to achieve a task goal (e.g. cats catching a mouse, or Orang-utans tying knots, Lestel et al., 2014) in his/her own way. It is important to recognise that behaviour is emergent under these interacting constraints (Chow, Davids, Hristovski, Araújo, & Passos, 2011). What this idea indicates is that expecting a prescribed outcome for each animal might limit individual capacity.

Environmental constraints are multilayered but most often presented as consisting of physical and sociological factors. Physical factors comprise the immediate surroundings and include physical influences such as gravity, altitude and the characteristics of behavioural contexts, such as ambient temperature, prevailing weather or whether the environment is familiar, novel, remote or physically demanding (Dillon et al., 2006; Paisley, Furman, Sibthorp, & Gookin, 2008). Sociological factors include the role of social contexts such as peer groups, and cultural expectations. In multispecies leisure the social context would include interspecies interactions where the human may not be at the heart of meaning making (Lestel et al., 2014). Social environmental constraints such as critical group members, the presence of support and access to high quality and appropriate infrastructure and facilities can have a powerful influence on behaviour (Kollmuss & Agyeman, 2002). In multispecies leisure research the human researcher might also be considered an environmental constraint. The skill, beliefs, fears, attitudes and research paradigm adopted by a researcher can have a positive or negative impact on an individual animal's behaviours. Rather than assuming that the researcher can adopt an objective stance this notion should be embraced and understood as part of the interactive process.

Task constraints consist of the goals of the specific task, conventions of the activity and the implements or equipment used during the experience. In contrast to the other constraints task constraints are easily manipulated, for example in learning contexts it might be important to consider interactive style or setting activities that are designed for different individuals. Due to non-proportionality and an appreciation that learning is nonlinear, small manipulations can often lead to large scale changes in an individual's behaviour.

Affordances

The concept of affordances is now well established within the fields of ecological and environmental psychology. Affordances are invitations for actions that stem from the relationship between an individual and the environment (Gibson, 1979; Sanders, 1997; Stoffregen, 2003). When an environment is described in terms of affordances the emphasis changes from a physical description to a functional description (Gibson, 1979). That is, the environment is described in term of what it offers the animal (for good or ill). For example, an apple affords eating for a hungry individual if that individual has the physiological structure to eat apples, the physical capacity to reach the apple and the apple is edible. This offers a shift away from a more traditional dualist view of the environment and individual and the notion of 'one-size fits all'. Instead it offers a rich framework that supports an individual approach to appreciating multispecies leisure. Gibson (1979) argued that the meaning of a particular environment for the animal (i.e. what it affords for action) is more relevant than physical qualities of the environment (i.e. structural qualities like colour or material, number or colour of other individuals and so forth), indicating that affordances might be shaped by individual, social or other influences, and that individuals express agency in selecting affordances from a landscape of opportunities for behavioural interaction. Perception and action of affordances is an embodied process where an agent actively interacts with their environment and acts upon the environment as the environment (social and physical) acts upon the agent in order to realise action possibilities. Visual perception, for instance, involves not just the eyes but the body and head as the agent moves to secure a more effective position for affordance realisation, as the use of blinkers to focus the attention of horses pulling carts and carriages illustrates. Perception and action of affordances also depends on goal directed intentions and individuals become attuned to information in the environment. For example, affordance perception and action might differ for the same individual in the same geographic area depending on whether the actor is intent on hunting, play, exploration or if other actors are around. In a social context when an animal is in the presence of another animal or animals the interrelated nature of the bidirectional influence is more obvious because the original animal experiences the other or others as living bodies (San Martín & Peñaranda, 2001), and therefore 'experiences' the other(s) as active agent(s).

In simple terms, a specific environment (social or physical) has specific properties that invite actions. For instance, affordances in the physical environment such as colour properties of water may be perceived as providing depth to dive into or shallowness to wade, and angles of inclination suggest different approaches to circumnavigation. Trees, for many, afford climbing opportunities, and gaps afford jumping across, stepping up and so on, depending on each individual's action capabilities. Equally, in the social environment a chair in full sunlight or warm human lap might afford relaxation for a cat (Lestel et al., 2014). Each individual perceives, utilises and shapes these opportunities for action from a unique perspective fashioned by their own individual constraints (Brymer et al., 2014). For example, two individuals climbing in trees might be working with the same environmental constraints but differences in individual constraints, such as limb length, body mass and previous experiences, could result in different opportunities for perceptions and actions. Objectively, a gap might be stepped over or leaped across and a tree might have climbing affordances but because of different individual constraints not all individuals can take advantage of the affordance. Affordances are dynamic and change as a function of time and context, illustrating the relevance of the person-environment scale of analysis. Different system states can influence the way that each individual interacts with the environment, for example, in constraining which affordances are perceived.

The implication of these ideas is that theoretical perspectives that focus on dualistic approaches and animals as species rather than individuals might be limited as tools for theoretical explanation. Instead, the ecological-phenomenological approach outlined here proposes that a relational understanding of multispecies leisure and animal-human interactions, where individual animals are perceived as embodied active agents, is a more effective medium for behavioural analysis (Fiskum & Jacobsen, 2012; Said, 2012). This idea emphasises that the mutual interaction between individual animals and their social and physical environment is key to interpreting multispecies interactions. The crucial idea is that the functional properties of the social and physical environment invite or encourage particular behaviours by providing ecological (i.e. task and environmental) constraints on animal behaviour.

An important relationship relevant to human-nonhuman interactions that is identified through the ecological approach concerns the animate-animate relationship (Gibson, 1979). That is the relationship between two perceiving, animate beings with 'Minds'. In this instance each animal is capable of perceiving the other and potentially capable of perceiving the other's Mind. In contrast to the Cartesian perspective, from an ecological-phenomenological perspective Mind does not exist solely in the head but is readily available to be perceived by other active agents with the capacity to perceive. In a context such as multispecies leisure this is important because all individuals will be distinguishing affordances for action that have direct meaning for themselves (Charles, 2011). However, differently from affordances in the physical environment, affordances in the animal-animal environment are dependent on the temporary properties of the animate individual who may be fearful, hungry, calm, satiated, asleep or receptive, for example (Gibson, 1979). Attuned individuals are able to perceive and respond to each other's mental states, goals, and intentions. What determines whether an affordance solicits action or how each individual experience affordances depends on individual constraints and whether the affordance has meaning for the animal. Further, the salience of affordances can be traced back to evolutionary niches. That is, not only might the same affordance solicit different actions for different individuals but some affordances might have greater solicitation 'power' based on evolutionary importance. In multispecies leisure this suggests the potential for conflicting actions, for example, while the immediate physical environment might seem fixed, affordances acted on might be different for each individual (Dings, 2018) depending on what it offers each agent at a particular point in time. An affordance might present itself as an attraction for one half of a multispecies pair in a particular leisure context, at a particular time, but as an avoidance for the other. In the following sections we highlight how this understanding might help develop a more meaningful interpretation of multispecies leisure.

Applying an ecological-phenomenological perspective to multispecies leisure events: horse riding pleasure rides

Horse riding is a popular form of multispecies leisure, and includes a variety of practices from trail riding to competitive sport. Trail riding, or 'hacking' as it is known in the UK, is a popular form of multispecies leisure that involves human and horse riding out together in open space – i.e. beyond the confines of an arena. Hacking can be a very enjoyable and relaxing experience for the rider, and many riders believe their horse also gains pleasure from the relative lack of structure and constraint placed on horse and rider as they traverse a variety of landscapes for exercise and enjoyment (Cochrane & Dashper, 2015). However, hacking can also be a stressful experience for horse and/or rider, as a variety of hazards may be encountered (from road traffic, to other animals, to other potentially scary sights, sounds and smells). The sense of freedom of hacking can also be a cause for concern for some riders, as beyond the relative safety of the arena some horses' behaviour can seem to change as he or she becomes more alert to her/his surroundings, which has the potential to result in the horse bolting – running off – with (or without) their rider. Consequently although often an enjoyable experience, some riders are fearful of hacking, particularly without the company of another horse and rider, and some horses also lack confidence out in open space (Dashper, 2017). As a result, increasing numbers of riders look to organised events called 'pleasure rides' to provide some structure and a sense of safety, whilst maintaining the enjoyable aspects of hacking. A pleasure ride is usually a one-day event that entails horses and riders following a predefined hacking route that is signposted and checked for accessibility and safety. Often run by equestrian charities as a fundraising activity, pleasure rides can see up to 100 horse-rider pairs set out on a route (usually between 10–20 miles) over the course of several hours.

The scenario

In this section we illustrate some of the potential benefits of applying an ecological-phenomenological framework to understanding hacking as a form of multispecies leisure in the context of a pleasure ride event. We begin by outlining a fictional pleasure ride scenario, developed from the first author's ethnographic research:

Participants:

- Human rider- female, mid 40s, 20+ years' experience of riding, had a hacking accident on previous horse 5 years ago, resulting in dislocated shoulder.
- Horse – gelding (castrated male), 8 years old, been a general riding horse for 4 years, partaking in hacking and jumping.
- The rider and horse have been a partnership for 2 years and hack out three times a week. They have attended four previous pleasure rides together.

Physical environment:

- A shallow river crossing, three quarters of the way round the route. The approach to the river is gently sloping and has pebbles. There are trees overhanging, which provide shade from the sun but may make the area appear dark.
- The river is not deep, varying from 20–40 cm, and horses have to cross a width of about three metres as part of the ride. Beyond the river is an open field.

This is a relatively common feature of pleasure rides, as in order to access countryside it may be necessary to cross water. However, not all horses are happy walking through a river, even if it is shallow, and may resist or even refuse to enter the river crossing. Pleasure ride event organisers provide participants with a map of the route, and will usually mark-up water crossings clearly to ensure riders are prepared. For some horse-rider pairs, this scenario may cause few problems and the horse may calmly walk down the slope and through the river crossing, before continuing on the ride. However, numerous issues may also arise. Sometimes the horse may spook on the approach to the river crossing, become scared or just resistant, backing up, spinning around and maybe even turning and running in the opposite direction. The rider may become concerned, tense up and communicate this corporeally to the horse, causing the horse to also become nervous and exacerbating any of these behaviours. An ecological-phenomenological perspective would be a useful tool for analysing this scenario and for understanding what is happening in this multispecies leisure encounter. This may provide pleasure ride event organisers with useful information on how individual horses may react to this kind of feature of a ride, which could inform future ride design and the production of supporting materials to help riders cope with any difficulties they may encounter in such a scenario.

An ecological-phenomenological analysis of the scenario

The first aspect to consider is that of constraints, beginning with individual constraints. The ecological-phenomenological approach offered here acknowledges that all animals – human and nonhuman – have individual constraints that are influenced by their species but not reducible to species differences. In the scenario above, the individual constraints of two actors need to be considered – horse and rider. Each will influence the unfolding of the situation. Some broad expectations can be made about how each is likely to behave in this encounter, but without knowledge of each individual animal it is difficult to determine intentionality and meaning and therefore predict what they will do and to understand why they behave as they do. For example, the rider is experienced and so might be expected to be able to handle this routine situation without much trouble. But she has had a serious injury, potentially as the result of a similar situation. Does she become more fearful as a result? Is she less clear than normal in communication with the horse? The horse also has some experience with hacking and pleasure rides, but some horses dislike water and are resistant to walking through rivers. Has this horse had a previous bad experience when walking through water? Does he have flatter feet, which makes walking on stony ground (the river crossing) uncomfortable? Is he getting tired on the ride, and so less willing to exert himself and communicate effectively with his rider?

The next analytical tool in this framework is to consider environmental constraints: physical and socio-cultural factors. In this scenario the physical environment plays an important role, setting out the space in which the encounter occurs and imposing limitations in terms of how horse and rider approach the river crossing, how they see the water and judge its depth, and how inviting this seems to both horse and human on this particular day and time. Other aspects of the physical environment may also come into play, such as the weather (which may affect visibility, or strong wind might make the horse more flighty), or if other horses can be seen in the opposite field cantering away, which may distract or excite the horse further. The socio-cultural environment also plays a role here. The rider is experienced and has been on previous pleasure ride events, and so she knows she should be able to get her horse past a relatively simple river crossing. However, she may be concerned about other riders on the event catching up with her and getting stuck behind her and the horse, or she may decide to wait for another rider to come along and use the horse's desire to be with other horses to her advantage so they can follow another horse through the water. The horse will also be aware that there are other, probably unknown, horses around and he may be more tuned in to them than his rider. He may want to rush through, or try to jump over, the water to catch up with horses in the next field, or he might want to turn back to other horses behind him for security. In this situation he may not be 'listening' to his human partner, and relying more on fellow horses to provide guidance and security.

The third type of constraint to consider relates to the task. It might be expected that a horse-rider pair that regularly hack out and attend pleasure ride events should be able to cross a simple river without too much trouble, but many horse-rider combinations do not often encounter water crossings in their routine riding activities, and so may have few opportunities to train for this scenario. The pair should be well enough attuned to communicate clearly with each other, but this can easily break down, or one or both of them may attune to unhelpful information in the environment during stressful situations, which the river crossing may represent for the horse and/or rider. Task constraints appear relatively easy to manage, but in association with individual and environmental constraints, outlined above, may prove to be less straightforward and predictable than expected, especially in a multispecies context involving multiple actors.

This seemingly simple situation offers a wide variety of affordances to each of the actors. For the horse, the physical environment invites several different, often contradictory, actions. The gentle slope and shallow water invite the horse to walk through it calmly and without much hesitation. However, if the horse dislikes water, or even if he had a bumpy, difficult journey to the event, the river might invite him to either jump it (potentially unseating the rider) or to refuse to enter it, leading to the horse becoming increasingly distressed and potentially even backing up and running away. The visibility of the open field beyond the river also invites the horse to rush through the river crossing, potentially unbalancing the rider, to get to an open space to gallop after other horses who can be seen in the distance. The situation also offers the rider many affordances. She can ride positively, quietly and effectively to guide the horse down the slope, through the water and calmly up the other side, reassuring him if he is nervous or calming him if he gets excited. However, if she had little sleep the night before or her previous riding accident occurred as the result of the horse spooking at water the situation invites her to behave differently, to become anxious, upset, and ineffective in her communications with the horse, potentially increasing the horse's distress and leading to a breakdown in interspecies communication.

A variety of outcomes may result from this scenario, dependent on the different actors involved, their interactions with each other and with the physical environment in which the encounter takes place. At a pleasure ride event, some horse-rider pairs will tackle the river crossing without incident, some may struggle to get through, and some may even result in the rider falling off and/or the horse bolting back to other horses or the event headquarters. The river crossing is a potentially hazardous aspect of the route for some horse-rider pairs, and consequently may result in negative feedback to the event organisers or even injury to horse or rider. The ecological-phenomenological approach outlined here offers analytical tools to try to understand how and why events unfold as they do, and why the different individual actors in the scenario behave in a certain way, which in turn can provide useful insight for event organisers designing pleasure ride routes and supporting materials such as maps, instructions and the provision of alternative routes, and situating stewards and first aid teams around the ride.

This brief discussion draws on a simple experience encountered during a pleasure ride to illustrate some of the complexities of multispecies leisure. Although a relatively mundane aspect of multispecies leisure, this scenario can develop in many different ways depending on the behaviours of both actors, and the influence of the wider environment on what unfolds. Our discussion above is a very simplistic application of some aspects of an ecological-phenomenological approach to this multispecies encounter, but it illustrates the relational aspects of such a situation, the complexities of multispecies interactions in leisure environments, and some of the challenges of understanding behaviours across individual and species barriers. We discuss some of these issues, and their practical, theoretical and methodological implications, further in the next section.

Discussion

The ecological-phenomenological approach to understanding multispecies leisure suggests that all animate actors in the leisure context, human and nonhuman, should be considered as individual embodied, active agents with their own set of intentions and individual constraints. Analytically this means that human animals are not privileged over nonhuman animals, as all actors are considered to have capacity to be attuned to environmental information and to influence the leisure act. In this way, the ecological-phenomenological approach attempts to decentre human experiences and interpretations, and thus begins to overcome deep-rooted human exceptionalism that characterises leisure studies (as well as all other social science subject areas). Within this perspective, human experiences are no longer privileged over those of other beings, and nonhuman animals are recognised as intentional, responsive, interpreting agents (Painter & Lotz, 2007) capable of meaning-making, acting and affecting actions and relationships within leisure spaces.

The ecological-phenomenological perspective we have introduced in this paper overcomes some of the potential limitations of ANT outlined previously in that it makes analytical space for individual animals and their unique abilities, corporeal realities and personal histories to shape leisure experiences. Whereas individual animals may disappear from focus in much ANT research (Cohen & Cohen, 2017), an ecological-phenomenological framework makes the individual(s), and their interactions with each other and the environment, the core focus of analysis. The fictional scenario presented above analysed with a conceptual framework informed by ANT may foreground other, nonanimate actors in this scenario, and consequently animate actors (both human and nonhuman) and their interactions with the environment may disappear from focus.

Human-animal studies, the other important more-than-human conceptual framework discussed previously, does tend to focus on individual animals – humans and nonhumans – and their interactions, but concentrates predominantly on specific human-nonhuman animal encounters and gives less analytical priority to the influence of wider environmental factors. The fictional scenario discussed above, analysed from a human-animal studies perspective, would focus on the relationship and dynamics between the rider and the horse, considering their past interactions and shared histories as much as the event itself. The ecological-phenomenological approach we have applied here moves away from specific interspecies relationships towards the relational aspects of encounters between individual animals (animate perceivers) and their environment. Multispecies leisure takes multiples forms, and involves a variety of different individual animals (human and nonhuman) in diverse leisure spaces, and the ecological-phenomenological approach enables researchers to consider both the importance of individual differences within each actor *and* how those differences interact with features of the leisure environment to impact and shape the leisure experience.

The ecological-phenomenological approach also indicates that research that attempts to consider multispecies leisure from the perspective of interpreting events from a species context will be limited in their capacity to draw meaningful conclusions. Within the example discussed in this paper, some broad predictions could be made about expected behaviour of both animate actors (human and horse) based on knowledge about general species behaviour, but this would provide limited information and understanding. Not all horses will respond in the same way to an environmental stimuli such as a river crossing, for example. Further, the same horse could behave differently on two different occasions. The ecological-phenomenological approach suggests that research requires in-depth knowledge of all actors within the multispecies leisure context, moving beyond generalisations at a species level. A full appreciation of individual, task and environment constraints, framed by context and time is also needed. This suggests that the researcher cannot adopt a neutral stance or approach multispecies leisure from an objective, detached position. Rather, the researcher needs to have knowledge and experience of the individual(s) involved, and preferably be intimately connected to the experience under consideration. In the example we discuss in this paper, the rider would be the ideal researcher, as she knows herself and the horse and can apply this understanding to the experience and how it felt. Ethnographic and autoethnographic methods may thus be ideally suited to this task. Researchers might also need to appreciate the importance of the concept of Mind being 'out there' and design methodologies that combine deep knowledge of individuals and animate-animate relationships. Equally methodologies that rely too heavily on snap shots that do not appreciate the importance of context or time or do not consider individual, environment and task constraints might at best be collecting partial data and at worst be collecting data that has no ecological validity. Interdisciplinary research, drawing on ethology and biology as well as social science perspectives, may offer deeper insights on individual behaviours, constraints, actions and affordances. As human-animal studies researchers acknowledge, to try to consider an encounter or an action from a nonhuman perspective is challenging, as we tend to revert to human interpretations, human frames of reference and human priorities in our research design, conduct, analysis and representation (Pacini-

Ketchabaw et al., 2016). Phenomenology, especially ecophenomenology, provides some guidance on how to interpret the lived experience of the 'other'. Rather than being fearful of our human perspective we should embrace this and open up to encountering the 'other' as a living body rather than an object of study (Ruonakoski, 2007). Effective research in multispecies leisure from this perspective is more likely to be useful if the researcher is attuned to the behavioural nuances of the 'other' and follows the phenomenological method, and maintains the phenomenological attitude (Brymer & Schweitzer, 2017). Just because it is challenging for researchers to move beyond humanism and try to understand the behaviours, actions and reactions of a nonhuman animal does not mean we should not try to do so if we truly want to appreciate a practice, such as leisure, as a multispecies encounter. Interdisciplinary research and innovative, flexible methodologies will be needed to try and achieve this.

The ecological-phenomenological perspective raises important theoretical and methodological issues in relation to multispecies leisure, but also has potential to inform practice. Event organisers and planners of multispecies leisure activities and facilities could adopt aspects of the approach outlined above to consider different issues that might arise at their event. The most important message from this approach is that one size does not fit all. Organisers need to be aware of the importance of animate-animate relationships and the key concepts that underpin this appreciation of multispecies leisure events. While it may not be possible to garner all information about possible individual constraints of partners in multispecies leisure it may be possible to gain a better appreciation of typical constraints or key affordances that might affect leisure outcomes. An appreciation of typical individual constraints might also help designers of multispecies leisure events and activities design tasks and environments with rich affordances for all animals, human and nonhuman, to facilitate safe, enjoyable and rewarding multispecies leisure experiences.

Conclusion

In recent years there has been a surge of interest in animal-animal relationships across a variety of scientific fields. This is also reflected in an appreciation of multispecies interactions in leisure contexts. This interest brings with it complications, particularly concerning methodology and how best to understand these relationships. The traditional perception of nonhuman animals that focuses purely on biology or a species approach has been criticised as too limited to be of any value to multispecies leisure research. In this article we proposed a new appreciation for this relationship predicated on understandings drawn from ecological psychology and phenomenology which recognise that the traditional subject-object dichotomy is flawed. From this perspective multispecies leisure involves various animate perceivers (human and nonhuman) in relationship with each other.

Methodologically, this suggests that the researcher needs to have intimate knowledge of the context and the actors involved. It is no longer appropriate to limit research to outdated object-subject dichotomies.

The ecological-phenomenological approach introduced in this paper has not previously been applied to research on multispecies leisure, but suggests exciting avenues for further research investigating animal agency and meaning in the leisure context, and the role of non-human animals in shaping the human leisure space. This approach might also provide a process for refining phenomenologically guided and other methodologies for investigating the varieties of non-human lived experience. This article is important, therefore, because it transforms current appreciation of multispecies leisure and opens doors to new ways of thinking and investigating the value and meaning of leisure in a multispecies context.

Disclosure statement

No potential conflict of interest was reported by the authors.

ORCID

Katherine Dashper 🆔 http://orcid.org/0000-0002-2415-2290
Eric Brymer 🆔 http://orcid.org/0000-0003-0274-1016

References

Badmington, N. (2003). Theorizing posthumanism. *Cultural Critique*, *53*(1), 10–27.
Beard, L., Scarles, C., & Tribe, J. (2016). Mess and method: Using ANT in tourism research. *Annals of Tourism Research*, *60*, 97–110.
Birke, L., & Hockenhull, J. (2015). Journeys together: Horses and humans in partnership. *Society & Animals*, *23*(1), 81–100.
Brymer, E., & Davids, K. (2013). Ecological dynamics as a theoretical framework for development of sustainable behaviours towards the environment. *Environmental Education Research*, *19*(1), 45–63.
Brymer, E., Davids, K., & Mallabon, E. (2014). Understanding the psychological health and well-being benefits of physical activity in nature: An ecological dynamics analysis. *Journal of Ecopsychology*, *6*(3), 189–197.
Brymer, E, & Schweitzer, R. (2017). *Phenomenology and the extreme sports experience*. Abingdon: Routledge.
Carr, N. (2014). *Dogs in the leisure experience*. Wallingford, UK: CABI.
Charles, E. (2011). Ecological psychology and social psychology: It is Holt, or nothing! *Integrative Psychological and Behavioral Science*, *45*(1), 132–153.
Charles, N. (2014). 'Animals just love you as you are': Experiencing kinship across the species barrier. *Sociology*, *48*(4), 715–730.
Charles, N., & Davies, C. A. (2008). My family and other animals: Pets as kin. *Sociological Research Online*, *13*(5), 1–14.
Chow, J.-Y., Davids, K., Hristovski, R., Araújo, D., & Passos, P. (2011). Nonlinear pedagogy: Learning design for self-organizing neurobiological systems. *New Ideas in Psychology*, *29*, 189–200.
Cochrane, J., & Dashper, K. (2015). Characteristics and needs of the leisure riding market in the UK. In S. Pickel Chevalier & R. Evans (Eds.), *Horse tourism and leisure: International scale – Local development* (pp. 82–91). Angers: Mondes du Tourisme.
Cohen, S. A., & Cohen, E. (2017). New directions in the sociology of tourism. *Current Issues in Tourism*, *2*, 1–20.
Costall, A. (2001). Darwin, ecological psychology, and the principle of animal-environment mutuality. *Psyke & Logos*, *22*, 473–484.
Dashper, K. (2017). *Human–animal relationships in equestrian sport and leisure*. Abingdon: Routledge.
Dashper, K. (2018). Moving beyond anthropocentrism in leisure research: Multispecies perspectives. *Annals of Leisure Research*. doi:10.1080/11745398.2018.1478738
Davids, K. W., Button, C., & Bennett, S. J. (2008). *Dynamics of skill acquisition: A constraints led approach*. Champaign, IL: Human Kinetics.
Dillon, J., Rickinson, M., Teamey, K., Morris, M., Choi, M. Y., Sanders, D., & Benefield, P. (2006). The value of outdoor learning: Evidence from research in the UK and elsewhere. *School Science Review*, *87*(320), 107–111.
Dings, R. (2018). Understanding phenomenological differences in how affordances solicit action. An exploration. *Phenomenology and the Cognitive Sciences*, *17*(4), 681–699.

Dowling, R., Lloyd, K., & Suchet-Pearson, S. (2017). Qualitative methods II: 'More-than-human' methodologies and/in praxis. *Progress in Human Geography*, *41*(6), 823–831.

Fiskum, T. A., & Jacobsen, K. (2012). Outdoor education gives fewer demands for action regulation and an increased variability of affordances. *Journal of Adventure Education & Outdoor Learning*, *1*, 1–24.

Gibson, J. J. (1979). *The ecological approach to visual perception*. Boston: Houghton Mifflin.

Gillespie, D. L., Leffler, A., & Lerner, E. (2002). If it weren't for my hobby, I'd have a life: Dog sports, serious leisure, and boundary negotiations. *Leisure Studies*, *21*(3–4), 285–304.

Haraway, D. J. (2008). *When species meet*. Minneapolis: University of Minnesota Press.

Immonen, T. J., Brymer, E., Davids, K., Liukkonen, J. O., & Jaakkola, T. T. (2018). An ecological conceptualization of extreme sports. *Frontiers in Psychology*, 9.

Kirksey, S. E., & Helmreich, S. (2010). The emergence of multispecies ethnography. *Cultural Anthropology*, *25*(4), 545–576.

Kollmuss, A., & Agyeman, J. (2002). Mind the gap: Why do people act environmentally and what are the barriers to pro-environmental behavior? *Environmental Education Research*, *8*(3), 239–260.

Lamers, M., van der Duim, R., & Spaargaren, G. (2017). The relevance of practice theories for tourism research. *Annals of Tourism Research*, *62*, 54–63.

Lestel, D., Bussolini, J., & Chrulew, M. (2014). The phenomenology of animal life. *Environmental Humanities*, *5*, 125–148.

Lloro-Bidart, T. (2017). A feminist posthumanist political ecology of education for theorizing human-animal relations/relationships. *Environmental Education Research*, *23*(1), 111–130.

Madden, R. (2014). Animals and the limits of ethnography. *Anthrozoös*, *27*(2), 279–293.

Maiteny, P. T. (2002). Mind in the gap: Summary of research exploring 'inner' influences on pro-sustainability learning and behaviour. *Environmental Education Research*, *8*(3), 299–306.

Newell, K. M. (1986). Constraints on the development of co-ordination. In M. G. Wade & H. T. A. Whiting (Eds.), *Motor development in children: Aspects of co-ordination and control* (pp. 341–360). Dodrech: Martinus Nijhoff.

Pacini-Ketchabaw, V., Taylor, A., & Blaise, M. (2016). Decentring the human in multispecies ethnographies. In N. Snaza & J. A. Weaver (Eds.), *Posthuman research practices in education* (pp. 149–167). London: Palgrave Macmillan.

Painter, C. M., & Lotz, C. (2007). *Phenomenology and the non-human animal: At the limits of experience*. Dordecht: Springer.

Paisley, K., Furman, N., Sibthorp, J., & Gookin, J. (2008). Student learning in outdoor education: A case study from the national outdoor leadership school. *Journal of Experiential Education*, *30*(3), 201–222.

Ruonakoski, E. (2007). Phenomenology and the study of animal behavior. In C. Painter & C. Lotz (Eds.), *Phenomenology and the non-human animal. Contributions to phenomenology (In cooperation with the center for advanced research in phenomenology)* (Vol. 56, pp. 75-84). Dordrecht: Springer.

Said, I. (2012). Affordances of nearby forest and orchard on children's performances. *Procedia–Social and Behavioral Sciences*, *38*, 195–203.

San Martín, J., & Peñaranda, L. M. (2001). Animal life and phenomenology. In S. Crowell, L. Embree, & S. J. Julian (Eds.), *The reach of reflection: Issues for phenomenology's second century* (pp. 342–363). Boca Raton: Center for Advanced Research in Phenomenology.

San Martin, J., & Peñaranda, M. L. P. (2001) Animal Life and Phenomenology. In Crowell, S., Embree, L., & S. J. Julián (Eds.), *The Reach of Reflexion: Issues for Phenomenology's Second Century* (pp. 342-363). Center for Advanced Research in Phenomenology, Inc. Electronically published at www.electronpress.com

Sanders, J. T. (1997). An ontology of affordances. *Ecological Psychology*, *9*, 97–112.

Sayes, E. (2014). Actor–network theory and methodology: Just what does it mean to say that nonhumans have agency? *Social Studies of Science*, *44*(1), 134–149.

Stoffregen, T. A. (2003). Affordances as properties of the animal-environment system. *Ecological Psychology*, *15*, 115–134.

Van der Duim, R. (2007). Tourismscapes an actor-network perspective. *Annals of Tourism Research*, *34*(4), 961–976.

van der Duim, R., Ren, C., & Jóhannesson, G. T. (2013). Ordering, materiality, and multiplicity: Enacting actor–network theory in tourism. *Tourist Studies*, *13*(1), 3–20.

van der Duim, R., Ren, C., & Jóhannesson, G. T. (2017). ANT: A decade of interfering with tourism. *Annals of Tourism Research*, *64*, 139–149.

Withagen, R., Araújo, D., & de Poel, H. J. (2017). Inviting affordances and agency. *New Ideas in Psychology*, *45*, 11-18. doi:10.1016/j.newideapsych.2016.12.002

Becoming horseboy(s) – human-horse relations and intersectionality in equiscapes

Eva Linghede

ABSTRACT

Leisure studies have given scant regard to human-animal relations and intersectionality. In this paper, I respond to calls for research analysing leisure as a complex, multispecies phenomenon by exploring human-horse relations and intersectionality in boy's/men's equestrian stories through the concept of *intra-activity* and creative analytical writing. Thinking and writing through intra-activity brings insights into the co-constitution of humans and horses, as well as the entanglement of other power relations and social categories. The paper illustrates that becoming horseboy(s) is a process of material-discursive intra-activity where boys/men, by transcending the human-animal divide simultaneously transcend the female-male/masculine-feminine divide. Thus, engaging materially with horses can allow and encourage boys/men to be less constrained by dominant gender discourses. The paper also illustrates the importance of studying gender, not as a separate or primary category of privilege or inequality, but as one that is entangled with race, class, sexuality, age and other animals. I finally argue that bringing horses, as well as discourses, into discussions of the enactment of gender in leisure landscapes offers a productive site for elaborating the much-debated question, posed by feminist posthumanists, of the agency of matter.

*

Something happens when I ask about the horses. Or when they speak about the horses themselves. Eyes that light up. Voices and bodies that soften. Warmth that trickles out between the lines. Affect. This is a dimension, maybe the only one, present in all interviews.
Field note 13/5 2012

In an earlier study, I interviewed 19 Swedish boys and men about their experiences of equestrian sports, a sport and leisure activity that in Sweden is heavily dominated by girls and women and commonly associated with 'girliness' (Hedenborg & Hedenborg White, 2012; Plymoth, 2012). One of the main conclusions was that participation in a female-coded leisure landscape, like equestrian sports, can open up for other, and less stereotypical, ways of being a man than participation in male-coded leisure landscapes (see Linghede, Larsson, & Redelius, 2015). By enacting that in equiscapes,[1] boys and men practice both masculine and feminine-coded positions, the *hi*stories challenged dominant discourses about men and women – how they are and what they like. Since I was interested in the construction of gender in boys and men's stories about equestrianism my analysis was guided by feminist poststructuralism, a theoretical framing that made me look for competing gender discourses and multiple positioning. However, already when thinking and writing the analysis, I felt that there was one aspect that I did not explore sufficiently: the relationship with the horses. As the field note above highlights, there was something with the horses, or the relationship with the horses, that seemed to *do* something with the boys and men. This was evident both during the interview sessions (in eyes, voices and body language) and when reading the interview transcripts. I started to think about how the relationship with another living being, like the horse, was involved in the construction of gender.

When later approaching posthumanist theorising and beginning to think with theorists like Karen Barad and Donna Haraway, I became even more curious to return to the interviews – now with a focus on the human-horse relation. Thinking with posthumanism means starting from the ontological assumption that humans are not separate, unified and autonomous subjects, but intimately entangled with other humans, animals, machines and the environment (Pyyhtinen, 2016). To speak with Haraway (1991) our bodies and those of others do not end at the skin but are rather active in naturecultural entanglements where anything present is potentially agential. In the growing field of human-animal studies, researchers emphasise a need to move beyond anthropocentric studies and explore how human and non-human lives are entangled and co-constitute each other. When it comes to horse-human relations there are studies enacting how horses become agentic in the self-construction as horse and human conjointly make each other up (see, for example, Birke, Bryld, & Lykke, 2004; Brandt, 2004; Dashper, 2017; Game, 2001and Maurstad & Davis, 2016), but as Maurstad and Davis (2016, p. 191) state 'there is a need for research that focuses on practices, relations and processes in order to increase understanding about how humans and horses grow as biosocial becomings'. Little research has also, as Birke and Brandt (2009) and Finkel and Danby (2018) point out, addressed questions of how gender is enacted within human-horse relationships.

Thinking back on the earlier interviews I also realise that I did not consider that the boys' and men's stories not only came from localisations marked and affected by gender and sexuality, but also by race, class, ethnicity and other matters. As intersectional researchers rightfully have pointed out, this analytical insensitivity to the entanglement of multiple categories and power structures has been symptomatic of gender studies in most disciplines (Collins & Bilge, 2016; Crenshaw, 1991; Davis, 2008). Leisure and sport studies are no exception and as Watson and Scraton (2013, p. 36) argue, 'thinking intersectionality is a useful means of analysing leisure as a dynamic interplay of individual expression and the social relations within which leisure occurs'. In retrospect, I can see that thinking through an intersectional lens, including more categories than gender and sexuality, would enrich my analysis, make it more complex.

Responding to calls for research exploring leisure as a complex, multispecies and intersectional phenomenon, I am, in this paper, doing a rereading of my earlier interviews. In this rereading or extended analysis, I explore both the human-horse relation and intersectional categories such as class, ethnicity/race and age. To my help, I 'plug in' Karen Barad's concept of intra-activity and use creative analytical writing as a method of inquiry. The paper is structured in three main parts. First, I introduce the concept of intra-activity and its methodological implications. Secondly, I illustrate how I have used dialogue as a method of inquiry. I then present the analysis in the form of two dialogues, or scenes, combined with more explicitly theoretical readings. Finally, I discuss how an exploration of human-horse relations and intersectionality as intra-action can contribute to the field of (gender) leisure studies.

Intra-action

Drawing on the works of among others Foucault and Butler, the philosopher and quantum physicist Karen Barad proposes a posthumanist notion of performativity – one that incorporates important material *and* discursive, human *and* non-human, and natural *and* cultural factors. For her, a posthumanist account 'calls into question the givenness of the differential categories of "human" and "non-human" examining the practices through which these differential boundaries are stabilized and destabilized' (Barad, 2003, p. 808). Barad uses the term material-discursive intra-activity as a way of crossing boundaries between nature and culture, non-humans and humans, and to enact how matter and meaning are always already co-constituted. The notion of *intra*-action (in contrast to the more common *inter*action) marks that it is not a question of interplay between independent and autonomous entities that meet but part unchanged, but rather of entanglements of inseparable entities in mutual transformation (Barad, 2003, p. 815). Referring to these entities as relata, Barad states that 'relata do not pre-exist relations; rather relata-within-phenomena emerge through specific intra-actions' (Barad, 2007, p. 140). The point is thus not merely that there are important material factors, for example, other animals, bacteria and hormones, in addition to discursive ones, rather, the issue is the co-joint material-discursive nature of constraints, conditions and practices (Barad, 2003, p. 823). Haraway (2008, p. 17) applies this thinking to human-animal relations, stating that 'The partners do not precede their relating; all that is, is the fruit of becoming with: those are the mantras of companion beings'. Hence, as partners that intra-act, both horses and humans change by engaging with each other. To explore the construction of gender in boys' and men's stories about equestrianism without considering how the horses are co-implicated in the formation of and enactment of gender, is, to speak with Barad (2003, p. 810), to cheat the horses out of the fullness of their capacity.

Intersectionality as intra-action

To explore how gender is enacted in human-horse relations one must also be attentive to other intra-acting dimensions, for example, how gender, sexuality, race, ethnicity, class, age and other matters are intertwined in and constitute each other. In kinship with Lykke (2010), I find it useful to think about intersectionality in terms of intra-action. One of the things which has been contested within an intersectional tradition is, namely, the usefulness of the very concept of intersectionality (Ken, 2007). Based on the metaphor of 'intersection', it implies that different roads cross at specific points but are otherwise unrelated. This is problematic if you hold the view, which most gender researchers do, that an intersectional perspective means conceptualising different categories and power dimensions as constitutive in the sense that they are not separate but entangled and mutually transformative (Collins & Bilge, 2016; Davis, 2008).

Thinking and writing with theory as methodology

From a Baradian perspective, we as researchers, do not obtain knowledge by standing outside of the world. Every I/Eye come from somewhere and it matters where we *site* and how we *sight* when we sense, analyse and (re)present the world; we are *of* the world, a part of its lively and ongoing intra-activity (Barad, 2003). Our ways of telling and naming (wording) is to create a world with material consequences (worlding) – for which we are responsible. Barad's onto-epistemology – where the observed object and the observer are not separated, but entangled – affects how qualitative researchers can engage with the world and has contributed to methodological reinventions, rethinkings and revisons of qualitative research practices. According to Giardina (2017, p. 265), such postqualitative practices are characterised by a turn towards thinking about ourselves more as entangled philosophers of inquiry rather than as researchers using fixed methods to gather data. What makes such 'thinking with theory' different from more traditional qualitative methods is that theory drives the methods and shapes how data and transcripts are produced, how one intra-acts with data and how one writes research (Kuby et al., 2015, p. 142). Writing is not something that happens after thinking and analysing but an intra-active part of research practice. To cite Richardson & Adams St. Pierre (2005), writing is a 'method of inquiry'.

In this paper, I think and write with Barad's concept of intra-activity. By using creative analytic writing, like dialogues and poetic pieces, I break with the depersonalised, decontextualised and bodiless writing style that is characteristic of traditional academic writing. I engender a position for experiencing myself as an intra-active sociological knower/constructor – not just talking about it, but doing it (Richardsson, 1992, p. 136). Creative analytical writing makes visible the intra-actions between the researcher and the researched and thus provides a way to expose and contest the constructedness of all writing. Even when research findings are represented in a more realist way, it is the researcher who chooses which quotations to exclude or include and give weight – in other words, which research story to tell (Richardson & Adams St. Pierre, 2005; Sparkes, Nilges, Swan, & Dowling, 2003).

Dialogue as a method of inquiry

By using dialogue as a method of inquiry I join a growing number of social science and leisure/sport researchers who make use of creative analytical writing to analyse and sensuously (re)present data that is already emotional, compelling and ambiguous and to evoke an empathic understanding with otherness (see for example Barker-Ruchti, 2008; Berbary & Johnson, 2012; Dowling, Garrett, Lisahunter, & Wrench, 2015; Sparkes, 1997; Sparkes et al., 2003). Starting from a premise of proximity rather than distance and recognising the role emotions play in our understanding of social phenomena, creative analytic practices like storytelling, poetry and dialogue can help us engage in and create other knowledge-worlds in a way that normative social scientific knowledge-making is less able to do (Dahl, 2012, p. 152; Richardson & Adams St. Pierre, 2005).

Interviews

The two dialogues in this paper are situated in the stories of 19 Swedish boys/men, between the ages of 13 and 55, who I interviewed about their experiences of equestrian sports in 2012. The boys/men were active on different levels and in different disciplines connected to the Swedish Equestrian Sports Federation. About half were competitive riders, some on the elite level, and about half were 'stableguys' who just liked to ride and hang around with the horses. Among the competitive riders, the disciplines show jumping, dressage, eventing, gymkhana, horse driving and endurance riding were represented. Ten of the interviewees were recruited through a list of names produced by the Swedish Equestrian Sports Federation and nine were recruited by going through lists on national team riders and members in local youth sections. In the selection of persons to contact I strived to get a variation regarding age, discipline, level and geographic residence. Of the boys/men, I finally interviewed all but one was white. No question was explicitly asked regarding sexual orientation, but several of the boys/men came out as straight during the interviews. One of the interviewees lived openly as gay. During the interviews, it also emerged that there were differences in class background.

The interview questions revolved around their way into equestrian sports, their experiences of equestrianism, treatment from others and views on women, men and equality. In line with the theory, I understand the boys' and men's' interview statements and stories as articulations of material-discursive constraints, conditions and practices, not as representations of essential subjects and experiences. Furthermore, the interviews were produced in intra-action with me, which means that I have to take my own location as a white, middle-class and female researcher seriously. Inhabiting categories such as 'white' and 'middle class' I run the risk of not paying enough attention to issues concerning, for example, race and class (Mohanty, 1988) – but also the responsibility for doing so (Lorde, 1984). Although I cannot give a full and final answer to the question of how my particular location influenced the knowledge production, I felt that differences and similarities in gender, age, and class – but also the fact that I was a researcher – mattered during the interviews. Sometimes I got the impression that the boys/men wanted to establish a good report with me as a woman (and gender scholar) by associating with positive attitudes about women and sometimes I felt that they, on the contrary, felt a need to underline 'macho' attitudes. I also think that they might have found it easier to discuss things like fears, feelings and fags with me as a female researcher. Some of the boys/men were inhibited by the interview situation and seemed to find it difficult to speak with a researcher in a relaxed manner. Perhaps this resembled the tension I felt when interviewing elite riders and boys/men with distinct 'macho' manners. The process of interviewing was, as always, played out in a complex web of power relations where the interview questions also contributed to a certain doing of gender. Presenting the analysis in dialogical form is a way of making my intra-actions with the interviewees, as well as with the transcribed interviews, visible.

Writing dialogues

The interview transcriptions were read and re-read with my eyes, ears and other senses on human-horse relations and intersectionality. Informed and inspired by the interview stories (as well as my bodily memories of the interview sessions) I then engendered a playwright/storyteller position from which I created two 'scenes', that could hold the identified intra-actions and display a contextual meaning (see Linghede et al., 2015 for a more thorough illustration of the process of transforming interviews into literary pieces). Based on the interview material, I created two dialogues written as an ongoing conversation between me and six characters (Adam, Bobby, Carl, Daniel, Eric and Filip), on a fictive thematic day about boys and equestrian sports. In the dialogues, and through the characters, a passionate venue was created between me and the interviewees, a meeting point where they and I, by reciprocity, got voice and body. To put it differently, the six characters you meet in the dialogues are created in intra-actions with theory, the interview material, and my imagination and knowledge of the field. Adam, Bobby, Carl, Daniel, Eric and Filip make visible ideas and thoughts rather than persons and show experiences and perspectives in a way that hopefully will reverberate with the reader not only as information, but also as emotions and desire. To highlight some theoretical ideas, I have also interwoven short, poetic vignettes in the dialogues.

Reading dialogues

Although the dialogues are both analytical and able to speak for themselves, I have chosen to include a more explicitly theoretical reading/discussion of them. In this reading, I follow and discuss some of the analytical threads enacted in the dialogues. However, I also want to invite readers to engage in meaning making from their own *sitings* and *sightings* – and potentially conspire to move beyond these. I thus want to open up for readings other and more than mine.

Scene 1: horseboy intra-actions

We're sitting in a conference room in a hotel in Stockholm, talking about horses and equestrian sports. Me and six boys/men. My colleague is in the room next door talking to seven others. They are all invited to the Swedish Equestrian Sports Federation's thematic day on 'boys and equestrian sports'. This morning conversations in focus groups is on the schedule. We've just been talking about their 'way into' equestrian sports and soon it's time for a coffee break.

Me: What would you say is the best thing about equestrian sports? What do you say Eric?

Eric: It's got to be all the girls! (*giggles*) No, to be honest, the atmosphere in the stable is MAGICAL. As soon as you get there it's like being in a different world. And the combination, the social aspect and the horses, the two most atmospheric things I know, and in the stable, I get them both.

Adam: For me it's the horses. Horses are AMAZING animals! After loving all the cute things about them, they just become more and more interesting. I can sit and look at them for hours.

And when you've known a horse for a long time, it becomes incredibly personal. It becomes a close, close friend. You can sense when it feels good, when it's happy, what it needs.

Eric: Yeah… horses are really special. They're silent friends who listen, and who love you for who you are. If you have a problem, you go to the stable and talk to a horse, bury your face in its mane. It just stands there, puts its head on your shoulder, maybe neighs. Then it's impossible not to get into a good mood.

Adam: The actual riding is also fantastic. The flow, the feeling when its going well, when everything just works. You're so in it, you don't think about yourself anymore, just about what you're doing right then and at the same time, instinctively, about what the horse is doing and how you're working together. Where does the horse end and I begin? When you really manage it it's GREAT! When you manage to think a thought through your body which is expressed by the horse and then you respond to that. But it can also be frustrating, when it doesn't work…

Carl: Yeah… but for me, I got hooked when I understood that it was about interaction and communication with the horse. I also played soccer and hockey before and then it was just me. If I ran faster and shot harder, I was better. But here that's not enough, you have to get the horse to work with you too.

I THINK about how their eyes light up. About the warmth in their voices. The softness. The humility. I think about the horses that DO something. The intra-action with the horses. Which shapes. Reshapes. Becoming with the horses. Becoming horseboy.

Eric: Ok, so I've never played soccer. I'm allergic to balls (*giggles*), but a horse is definitely not a ball that you kick around and then put back in the cupboard when you're done. Or buy a new one if it breaks. It's more like a family member.

Bobby: Yeah, my old horse, Bilbo, he had SO many feelings, he was like a human. He taught me that you can't just hop onto the horse and ride. You have to build a relationship first. And at the same time, that's what makes it so hard if a horse has to be sold or gets hurt or has to be put down. It's so damn hard.

Filip: It's awful. I don't know if you've seen that movie War Horse? I saw it the other day and cried my eyes out. It portrays the relationship between a horse and a human in such a beautiful way. Just the way it is, you pour in your whole soul, and when you see your old horse again and you recognise each other, that feeling, it never goes away…

Daniel: I think that's the reason I stopped riding for a while. I didn't have the energy to start again. But then I found a horse that was neglected, not my type at all actually, and too small also. But I couldn't just leave it there, so I bought it even though I didn't even have a place in a stable for it! I worked my ass off just to make it work. I mean, that shows what kind of commitment it's about.

Carl: But all that time that you spend on the horses, I think it teaches you a lot too. Being responsible for another being, making sure that it's well and has everything it needs. You always have this animal that needs food at certain times and that needs to be ridden and taken care of, you can't just ignore that.

Daniel: And maybe a horse suddenly gets colic and you have to walk with it for four hours. Then you can't just put it back after an hour and say that it's fine because you have to go home and do something else. You'll just have to change your plans.

I THINK about all the emotions. About the care. And sorrow. I think about the horses that DO something. The intra-action with the horses. Which shapes. Reshapes. Becoming with the horses. Becoming horseboy.

Me: It's interesting to listen to your conversation in relation to some studies that I've read. Because in those I've read that these aspects, caring for the horses, grooming them and cuddling with them, is something that mostly girls and women appreciate and spend time on. But I see you light up, and there's such warmth and passion in your voices when you talk about your relationship with the horses, and your closeness to them. What do you think about that?

Carl: Ok, well, it's like this. The interaction with the horses is amazing. There's nothing that can beat the feeling when you and the horse really find each other. The freedom when you ride. But when I was young I didn't enjoy hanging out in the stable at all. I mean, brushing the horse until it shines and competing in grooming competitions, what guy likes that? But then you have to take care of the horse in order for it to feel good and not chafe and stuff like that, but that's different.

Eric: But I love taking care of the horses and cuddling with them! *(blushes)* AND competing in grooming competitions. The first thing I do when I get to the stable is to go to the horse that I'm the carer for and give him some extra attention, check how he's doing and stuff like that.

Carl: Ok *(nervous laughter)*. Yeah, maybe I shouldn't speak on behalf of all guys. But honestly, I think MOST guys would agree with me.

Adam: Damn, you're brave Erik. I wouldn't have dared to say that when I was 14 years old. Even if I felt the same way. And still do.

Bobby: For me it depends. I mean, of course I cuddle with them, but I think that if you want to get something done, then it's all about serious training. But when you get to the paddock in the morning and they're all sleeping, and their bellies are their highest point, then you just want to go out and sit with them and cuddle. It's super cosy.

Daniel: One thing doesn't have to exclude the other, does it? But also, horses are big animals. I mean, it takes a certain psyche to handle a 600 kg horse. But it's a really powerful feeling when you feel that you can handle a stallion at Flyinge, that you can lead it back and forth to the stallion pen.

Bobby: In that way it's a bit weird that equestrian sport is seen as a girl's sport. Handling horses is hard work, and it's a tough and pretty dangerous sport.

Carl: Yeah, I've broken both arms and legs, and my nose *(laughter)*. I've definitely gotten more injuries riding than playing soccer or ice-hockey, that's for sure.

Eric: Also, if you look at pictures from the First World War for example, then the only people who were riding were men, and it was very 'macho'. Imagine... it was these big sweaty soldiers who were grooming the horses and scraping their hooves.

Filip: I mean, yeah, sure you have to be determined and strong. But at the same time, you need empathy to deal with animals. You have to be able to relate to how the animal acts and feels, otherwise you can't cooperate. And empathy isn't usually associated with masculinity and high levels of testosterone *(laughter)*. In some way I think that matters, that you need to be responsive and take the horse into consideration, you can't just steamroll over it.

Daniel: And you also have to be able to let go of some control. Otherwise there's no interplay. Maybe you don't have to 'surrender to the animal' like some people say, but you have to be able to let go, release some control. I think some guys are a bit afraid of that.

Carl: Speaking of control, weren't we supposed to have a coffee break at 10? It's five minutes past now and I'm really craving for coffee.

I THINK about norms. About gender norms. I think about the horses that DO something. The intra-action with the horses. And the norms. Which shape. Reshape. Becoming with the horses. And discourses. Becoming horseboy.

A reading

A common thread running through the dialogue is that the boys/men relate to dominant gender discourses. They know that equestrian sport is considered a 'girls sport' and that as boys/men, they are expected to be strong, brave, controlled, rational, competitive, and attracted to girls/women. Another, and entangled, thread is the close, mutual and intensely physical human-horse partnership that has evolved over the years they have engaged in horse care, riding and training. To speak with Birke et al. (2004), what characterises the human-horse relationships is that both horse and human must learn how to participate in a co-joint world where neither horse nor human emerges as a pre-existing category but as something produced by their co-joint actions. Thus, to explore how gender is performed in equiscapes one must include the actions of the horse. Humans and horses are engaged in a kind of choreography in which gender becomes an accomplishment of both horse and human. In this paper, which is based on interviews, I focus on the becoming of boys/men in intra-action with horses. However, it would be equally interesting to explore the (gendered) becoming of horses in intra-actions with humans.

Taking a closer look at some of the intra-actions, focusing on what the horses do, the boys/men in the dialogue describe that engaging with horses, when riding and when on the ground, allow, encourage, influence and render possible a set of human actions. On the one hand, they have to be(come) tough, capable and determined, features that are usually associated with masculinity. On the other hand, and perhaps above all, they have to be(come) communicative, sensitive, responsive, emphatic and caring, features that are more associated with femininity. For the boys/men figuring in the dialogue horses are both soulmates and bodymates, and as Filip puts it, such emotional and emphatic relations are not directly associated with 'high levels of testosterone'. To explore this intra-action further, the practices of training, learning to ride, riding with and being in the company of horses comprises a form of embodied mutuality where non-verbal forms of communication are at the centre. When handling horses on the ground, body language and touch are essential parts of the human-horse communication. To speak with Sedgewick 'touch ramifies and shapes accountability. Accountability, caring for, being affected, and entering responsibility are not ethical abstractions; these mundane, prosaic things are the result of having to truck with each other'. Furthermore, touch increases the level of oxytocin (a hormone connected to a sense of calm, decreased anxiety, protection against stress, increased trust, and empathy) in both humans and other animals (Kuchinkas, 2009). The two bodies of human and horse are thus, through touch, intra-acting down to the level of molecules.

Moving on to riding, body-to-body communication is at the heart of interaction (Dashper, 2017, p. 31). As Adam points out, riding is a process that requires extreme sensitivity on the part of both rider and horse. Game (2001) describes the mutually transformative process of riding as 'embodying the centaur' and Dashper (2017, pp. 70–90) employs the notion of 'feel' to explore the shared sensorial communication that shapes and reshapes both horse and rider. Here I want to chime in that I view the human-horse relation as mutual, but not equal. I thus agree on the importance of paying attention to the unequal power structures structuring the human-horse relation (as well as relations between humans). However, I also believe that highlighting the agency of the horse, and to turn from dualism to intra-action, is an affirmative way of questioning the same power differences.

I argue that the dialogue illustrates that becoming horseboy is a process of material-discursive intra-activity where engaging materially with horses allow and encourage boys/men to be less constrained by dominant gender discourses. In other words, by transcending the human-animal divide, the boys/men simultaneously transcend the idea of a female-male or masculine-feminine divide. Others have shown that the same could be said for girls and women. Since horse care, training and riding allow and require girls/women to be tough, competent, strong, capable and not mind getting dirty it creates and provides opportunities for them to challenge normative gender ideals too (Birke & Brandt, 2009; Dashper, 2016; Finkel & Danby, 2018).

Scene 2: intersectionality as intra-action

We've fetched coffee and sit down by one of the round tables in the lobby outside the conference room. The intimate atmosphere that arose when we talked about the horses suddenly feels far away.

The conversation is slow now that there aren't any specific questions to talk about. I decide that it's time to start with the next area of questions.

Me: But what's it been like to be a guy in such a predominantly female sport?

Eric: Very positive (*smiles*). I mean, people are always so happy when there's a guy, since there's so few of us. I get a lot of extra attention in the stables, and maybe some extra perks too (*laughs*). And since I'm not socially afraid, a get a lot of cool opportunities.

I see that several of the others nod in agreement

Carl: Now, I'm a very competitive person, but I think that the fact the I got a lot of positive feedback in the beginning also had an impact. I had a really good riding instructor who appreciated that there were guys and gave some extra support.

Bobby: As a guy you automatically become that red piece of paper in a pile of white papers. That's definitely an advantage. I mean, if you win all the time maybe it doesn't matter, then people know who you are no matter if you're a guy or a girl. But otherwise you pay a bit more attention to that guy in the field, since he's one out of thirty. I definitely think that I've gotten the opportunity to ride as many horses as I have because I'm a guy. People have heard of you, seen you in competitions, and then they come up and talk to you...

Me: So, WITHIN equestrian sports it's mainly positive to be a guy? Or are there any negative aspects?

Filip: Well, it can add pressure too. To be that red piece of paper that Bobby talked about. If you screw up too many times you become a neon-colored paper, and that's just uncomfortable to watch. (*laughter*)

Adam: I think you can get TOO much attention. The instructors that I had in the beginning gave me more feedback during the lessons than they gave to the girls in the group, which made the girls sad and annoyed. In the end I had to tell them that I didn't want any special treatment.

Daniel: The only negative thing I can think of is that I got a bit put down by the girls in the stables sometimes. They were on me and said I wasn't doing things right when I was grooming and stuff... But on the other hand, that became an incentive when I started competing. I wanted to prove that I was good at something. Generally, I think that if you're a guy and you show that you´re talented, you get noticed for that. At least if you are into dressage or show jumping, which is most common. In other disciplines, like in racing, it might be different...

I think about norms. Gender norms. I think about the fact that horsegirls and horseboys BECOME different. Are made different. In intra-actions with people and horses. Becoming with discourses. And horses. Becoming horseboy(s).

Me: What's it been like outside the sport? In school and other contexts?

Daniel: In school it wasn't exactly great. I was pretty bullied. Especially in elementary school and junior high.

Adam: I'm from a small town and have always attended schools with a lot of 'sports divas', soccer guys, and they're the worst. The only thing that got me through it was the fact that I knew that the shit would end at two or three in the afternoon. When I went to the stables. To the horses and my friends there.

Me: So, what could people say to you in school?

Adam: For example, they used to neigh when I walked past, or ask where I had my riding pants. And they used to call me things like 'horseboy', 'girl' and 'horse fag'. Stuff like that.

Everyone except Eric nods in agreement. He looks surprised.

Adam: When I started high school, I didn't tell anyone I did horse riding until after a - whole year. I just couldn't handle it. And then they were like, 'but you're a normal person!'. They got to see that you could be into horses and still be a normal guy.

Me: And what's a normal guy like?

Adam: Wow, I sounded really prejudiced when I said that. Well... eh... not gay then. That's what people think, I mean. That guys who do horse riding are gay.

Filip: Yeah, I got called fag when I was just eight years old. But I didn't become one until I was a teenager.

Eric: I've almost never been teased because I do horse riding. My classmates think it's cool. At least as far as I know.

Bobby: It's like that for me too. NOWDAYS. But I got harassed and teased all the time until people found out that I'd won the national championships. That was sometime in junior high school I think. Then it was like the guys suddenly realised that I wasn't just doing something silly. Now I almost think they're a bit jealous of me. Especially since I get to hang out with so many good-looking girls (*blushes*).

Carl: Hey, if your classmates think it's cool Eric, I congratulate you. But I wander what kind of school you actually go to? (*rolls his eyes and smiles*) For me it didn't change until my twenties. Then all of a sudden, a lot of people were like 'Damn, that's cool. I wish I would have tried that'. But before that you'd pretty much have to hold a gun to my head to get me to confess to doing horse riding, even though I was competing successfully from the age of 15. I grew up in Biskopsgården[2] and if you told people there that you did horse riding, maybe especially if you're black, you'd get your ass kicked. I even stopped riding for two years when I was 12 because I didn't want to be a geek.

Filip: Yeah, something happens when you get older. If that's any comfort. I was at a party last weekend, with a bunch of new people, and they couldn't stop talking about horses (*laugher*). They were SO interested and fascinated. Both the guys and the girls.

I THINK about Eric's well-fitting shirt and briefcase. About his self-evident way of taking place, talk and move. 14 years old and SO self-evident. I think about Carl smiling ironically and rolling his eyes when asking what kind of school Eric goes to. I think about Carls macho-manners. How he stresses that boys and girls are different and how he downplays what is 'girly'. I think about class, skin-color, and ethnicity. I think about age, gender, and sexuality. I think about intersectionality as intra-action.

Me: The fact that you get called gay, why do think that is?

Adam: I think it's because a lot of people think that horse riding is a girl's sport, and if you're a guy who does horse riding then you must be a bit girly, and therefore gay. Or very rich. So, either gay or a snob.

Carl: But its's also not completely wrong. There are quite a lot of gay guys in equestrian sports. I'm not sure why. In a way it's a very tolerant and open sport. At least I've never heard that anyone cares that there are a lot of gay people.

Adam: I think it's easier to be open since there are so many girls. It's not a macho-sport, like soccer or hockey.

Daniel: I don't think I'd known as many gay people if I hadn't been a riding instructor. But this thing that people think you're gay if you do horse riding, I mean since I'm..... No, never mind, I'm just making it complicated.

Me: No, go on, since you....?

Daniel: Well, since I'm not gay myself I've always had this kind of uncomfortable feeling that people are going to treat me like something I'm not (*draws breath*) ... I mean like I'm gay. And that's kind of stuck with me. I think it's a bit uncomfortable to see stuff on TV about gay people and stuff like that.

Carl: Wow, for me it's the complete opposite. I think I would've had more prejudices if I hadn't done horse riding. Also, it's a really good environment for us who don't swing that way, because of all the girls (*laughter*).

Filip: I don't agree that it's easy to be openly gay in equestrian sports. I only felt that I could be completely open when I wasn't competing on an elite level anymore. Now, I haven't got anything to lose anymore. I don't need anyone's approval. But I still have the feeling that people think my openness is provocative. On the other hand, that's hard to know because no one says anything to your face. And I've actually thought a lot about this. Why is it hard to be open in a sport that's seen as a gay-sport? And Daniel, I don't think you're the only one who feels like you do. Unfortunately. And that's kind of the thing. Since people think that if you're a guy who does horse riding, you're gay, it becomes extra important to show that you're not. And that's really sad. Cause what does that say about what you think about gay men.

*I THINK about gender and sexuality norms. About how they intra-act with each other and Other. Norms. I think about the word **or**. About Adam saying that if you are a gay who do horseriding people think you are either gay **or** snob. Imagining that you could be **both** seems too difficult. I think about intersectionality as intra-action. And about becoming(s). With horses and discourses.*

A reading

For me, to think intersectionality is to momentarily disentangle a complex knot of intertwined power relations and social categories. Since different power relations and social categories are actualised in different situations, the boys/men figuring in the dialogue are responding to a meshwork of intra-acting discourses – as well as to the horses. Starting off in the entanglement of gender and sexuality discourses, the boys and men figuring in the dialogue enact that engaging in a leisure activity considered feminine or girly, you run the risk of getting your (hetero)sexuality questioned. One way of handling this is to emphasise masculinity – especially heterosexuality – to show that you, after all, are a normal guy. As Filip, who lives openly as gay, puts it: *Since people think that if you're a guy who does horse riding, you're gay, it becomes extra important to show that you're not.* In my reading the fag becomes a border agent – a stereotype that both restricts and allows (after all, he offers a thinkable position to inhabit), but above all makes visible a heteronormative order in which gender and sexuality norms are deeply intertwined.

Gender and sexuality were the categories I, inspired by feminist poststructuralist and queer theory, focused on in the initial study; a theoretical siting that may, in turn, have been chosen because of my intersectional location as a white, middle-class woman living as a lesbian. This is not to say that exploring the entanglement of gender and sexuality is not relevant. In kinship with Richardson and Adams St. Pierre (2005), I argue that having partial knowledge is still knowing. However, adopting an analytical multi-sensitivity, taking into account that people, including myself, come from localisation also marked by race/ethnicity, class and age, extends the analysis, make it more complex. To speak with Davis (2008., p. 77) 'with new intersections, new connections emerge, and previously hidden exclusions come to light'.

Returning to the dialogue and the resistance the boys/men have met outside equestrianism, because of them engaging in a 'girly' or 'gay' sport, their experiences differ somewhat. For example, Eric, whose clothes and manners signal upper-class, stand out by saying that his classmates have never teased him, but rather think it is cool that he is doing horse riding, while Carl, who is black and has grown up in a socio-economically deprived area, just could not confess he was doing horse riding since it was the same as getting your ass kicked. For Adam, who lives in a small town and has gone to school with a lot of 'sport divas' it was tough in school, although he found a resting place among the girls and horses in the stable. Thinking about these differences through an intersectional lens, I argue that they enact how class, race/ethnicity and context intra-act with gender (and sexuality) norms. The expectations on how to live and act as a boy/man differ depending on your intersectional location. Due to the historical connections between equestrian sports and the upper-class (Dashper, 2017, p. 7), it is probably more accepted (and sometimes even expected) to engage in horse riding if you are a boy born and bred in an upper-class environment. This could help explain Eric's unilaterally positive experiences. I also read Carls ironic smile and the rolling of his eyes when asking what kind of school Eric goes to, as a hint towards this upper-class background. In turn, Adam's and Carl's statements could be read as expressions of their upbringing in sporting and working-class environments. These are contexts that often are characterised by a hegemonic masculinity ideal, where approaching female-coded practices are heavily sanctioned (Messner, 1995; Willis, 1977). Also, Berggren (2013) illustrates that in male resistance against racialisation, stereotypical discourses on gender and sexuality are often drawn on. While combating racism, the norm that masculinity consists of heterosexuality, distancing from the feminine and none cowardness is reproduced and reinforced. Thus, a possible reading of Carl's macho-manners and downplaying of grooming and cuddling with horses is that it has to do with his intersectional location as a black man brought up in a deprived area engaged in horse riding. However, apart from Carl quickly mentioning that he is black, race ore skin-colour is not, unlike, for example, gender and sexuality, explicitly discussed in the dialogues (neither by me nor the boys/men). This is a silence that enacts how whiteness is taken for granted and how the absence of bodies of colour does not give rise to feelings loss or lack in white contexts (Moraga & Anzaldúa, 1983).

The boys/men in the dialogue also enact that gender intra-acts with age. Doing horse riding is problematic for boys in their teens, but as they grow older and gender expectations change, it rather gives them appreciation. Furthermore, I argue that thinking through an intersectional lens, it is also important to take the local context into account. Regardless of your position in an intersectional web, you can be privileged in one local context and disadvantaged in another. For example, the boys/men in the dialogue describe that they have met a lot of resistance *outside* equestrian sports. However, *within* equestrianism, they have mainly had a privileged position.

My reading of the dialogue thus illustrates the importance of studying gender, not as a separate or primary category of privilege or inequality, but as one that is entangled with race, class, sexuality, age and local context – and other animals. These categories cannot be added or subtracted from each other at will but are mutually constitutive.

Discussion

Leisure studies have given scant regard to human-animal relations and intersectionality. By exploring human-horse relations and intersectionality in leisure *as intra-action* this paper brings insights into the co-constitution of humans and animals and various other power relations and social categories. In the paper, I illustrate how human-horse engagements are meshworks; complex interweaved series of intra-action that shapes and reshapes both humans and horses. The shift from interaction to intra-action accomplish a move from states of being where subjects and things are discrete with given essences to movement, relations and processes where subjectivities are not fixed, but develop and take shape through multispecies and multicategorical encounters. The notion of intra-activity is thus a way of challenging the separation of humans from animals, nature from culture, material from discursive, male from female, researcher from researched and theory from method; dualisms that from a posthumanist onto-epistemological stance is a remnant from anthropocentric Enlightenment ideas (Barad, 2003). As the dialogues in this paper enact, becoming horseboy(s) is a complex, multispecies and intersectional process in which a transcendence of the human-animal divide simultaneously allows and encourages a transcendence of the female-male/masculine-feminine divide.

I also argue, that bringing horses (as well as discourses) into discussions of gender performativity in leisure offers a productive site for elaborating the much-debated question, posed by feminist posthumanists, of the agency of matter (see, for example, Barad, 2003, 2007; Grosz, 1994; Haraway, 1991). As Barad puts it, 'any robust theory of the materialization of bodies would necessarily take account of how the body's materiality – for example, its anatomy and physiology – and other material forces actively matter to the processes of materialization' (2003, p. 809). In kinship with Birke et al. (2004), I argue that since animals are less easily discarded than human bodies and non-living matter as agents in their own right, they might offer a productive entrance into the discussions about how matter matters – or rather how matter and meaning are always already co-constituted.

Notes

1. Finkel and Danby (2018) describe equiscapes as a horse-focused leisure landscape where interconnectedness, emotional exchange and cross-species communication are encouraged. Hence, boundaries between human bodies and horses become blurred and entangled.
2. Biskopsgården is a socio-economically deprived area in Gothenburg.

Disclosure statement

No potential conflict of interest was reported by the author.

References

Barad, K. (2003). Posthumanist performativity: Toward an understanding of wow matter comes to matter. *Signs*, *28a*(3), 801–831.

Barad, K. (2007). *Meeting the universe halfway. Quantum physics and the entanglement of matter and meaning*. Durham: Duke University Press.

Barker-Ruchti, N. (2008). "They must be working hard": An (Auto-)ethnographic account of women´s artistic gymnastics. *Cultural Studies – Critical Methodologies, 8*, 372–380.

Berbary, L., & Johnson, C. (2012). The American sorority girl recast: An ethnographic screenplay of leisure in context. *Leisure/Loisir, 36*(4), 243–268.

Berggren, K. (2013). Degrees of intersectionality: Male rap artists in Sweden negotiating class, race and gender. *Culture Unbound, 5*, 189–211.

Birke, L., & Brandt, K. (2009). Mutual corporeality: Gender and human/horse relationships. *Womens Studies International Forum, 32*, 331–347.

Birke, L., Bryld, M., & Lykke, N. (2004). Animal performances. An exploration of intersections between feminist science studies and studies of human/animal relationships. *Feminist Theory, 5*(2), 167–183.

Brandt, K. (2004). A language of their own: An interactionist approach to human-horse communication. *Society & Animals, 12*(4), 299–316.

Collins, P., & Bilge, S. (2016). *Intersectionality*. Cambridge & Malden: Polity Press.

Crenshaw, K. (1991). Mapping the margins. Intersectionality identity politics and violence against women of colour. *Stanford Law Review, 43*(6), 1241–1299.

Dahl, U. (2012). The road to writing. An ethno(bio)graphic memoir. In M. Livholts (Ed.), *Emergent writing methodologies in feminist studies* (pp. 148–165)). New York: Routledge.

Dashper, K. (2016). Strong, active women: (Re)doing rural femininity through equestrian sport and leisure. *Ethnography, 17*(3), 350–368.

Dashper, K. (2017). *Human-animal relationships in equestrian sport and leisure*. New York: Routledge.

Davis, K. (2008). Intersectionality as a buzzword: A sociology of science perspective on what makes a feminist theory successful. *Feminist Theory, 9*(1), 67–85.

Dowling, F., Garrett, R., Lisahunter,, & Wrench, A. (2015). Narrative inquiry in physical education research: The story so far and its future promise. *Sport, Education & Society*

Finkel, R., & Danby, P. (2018). Legitimizing leisure experiences as emotional work: A post humanist approach to gendered equine encounters. *Gender Work & Organization*, 1–15.

Game, A. (2001). Riding: Embodying the centaur. *Body & Society*, *7*(4), 1–12.

Giardina, M. (2017). (Post?)Qualitative inquiry in sport, exercise, and health: Notes on a methodologically contested present. *Qualitative Research in Sport, Exercise and Health*, *9*(2), 258–270.

Grosz, E. (1994). *Volatile bodies: Toward a corporeal feminism*. Bloomington: Indiana University Press.

Haraway, D. (1991). *Simians, cyborgs and women: The reinvention of nature*. New York: Routledge.

Haraway, D. (2008). *When species meet*. Minneapolis: University of Minnesota Press.

Hedenborg, S., & Hedenborg White, M. (2012). Changes and variations in patterns of gender relations in equestrian sports during the second half of the twentieth century. *Sport in Society*, *15*(3), 302–319.

Ken, I. (2007). Race-class-gender theory: An image(ry) problem. *Gender Issues*, *24*(2), 1–20.

Kuby, C. R., Aguayo, R. C., Holloway, N., Mulligan, J., Shear, S. B., & Ward, A. (2015). Teaching, troubling, transgressing: Thinking with theory in a post-qualitative inquiry course. *Qualitative Inquiry*, *22*(2), 140–148.

Kuchinkas, S. (2009). *Chemistry of connection: How the oxytocin response can help you find trust, intimacy, and love*. Oakland: New Harbinger Publications.

Linghede, E., Larsson, H., & Redelius, K. (2015). (Re)presenting equestrian *histories* – Storytelling as a method of inquiry. *Sport, Education and Society*, *21*, 82–95.

Lorde, A. (1984). *Sister outsider: Essays and speeches*. Trumansburg: Crossing Press.

Lykke, N. (2010). *Feminist studies: A guide to intersectional theory, methodology and writing*. New York: Routledge.

Maurstad, A., & Davis, D. (Eds). (2016). *The meaning of horses: Biosocial encounters*. London: Routledge.

Messner, M. (1995). *Power at play: Sports and the problem of masculinity*. Boston: Beacon Press.

Mohanty, C. (1988). Under Western eyes: Feminist scholarship and colonial discourses. *Feminist Review*, *30*, 61–88.

Moraga, C., & Anzaldúa, G. (1983). *This bridge called my back. Writings by radical women of colour*. New York: Kitchen Table.

Plymoth, B. (2012). Gender in equestrian sports: An issue of difference and equality. *Sport in Society*, *15*(3), 335–348.

Pyyhtinen, O. (2016). *More-than-human sociology: A new sociological imagination*. Houndmills: Palgrave Macmillan.

Richardson, L., & Adams St. Pierre, E. (2005). Writing. A method of inquiry. In N. Denzin & Y. Lincoln (Eds.), *The Sage handbook of qualitative research* (3rd ed., pp. 959–978). London: Sage Publications.

Richardsson, L. (1992). The consequenses of poetic representation: Writing on the other, rewriting the self. In C. Ellis & M. Flathery (Eds.), *Windows on lived experience* (pp. 125–140). Newbury Park, CA: Sage.

Sparkes, A. (1997). Ethnographic fiction and representing the absent other. *Sport, Education and Society*, *2*, 25–40.

Sparkes, A., Nilges, L., Swan, P., & Dowling, F. (2003). Poetic representations in sport and physical education: Insider perspectives 1. *Sport, Education and Society*, *8*, 153–177.

Watson, B., & Scraton, S. (2013). Leisure studies and intersectionality. *Leisure Studies*, *32*(1), 35–47.

Willis, P. (1977). *Learning to labour: How working class kids get working class jobs*. Hampshire: Aschgate.

An exploratory study of British Millennials' attitudes to the use of live animals in events

Elena Marinova and Dorothy Fox

ABSTRACT

Ethical issues related to animal rights have gained significant exposure in the past few decades. As a result, animal welfare concerns have continuously been at the forefront of public debate. This has had a major impact on Western culture, expressed in the growing popularity of lifestyle changes towards reducing and abandonment of animal use across different industries. However, animal use in planned events remains insufficiently studied and absent from most event management literature. Therefore, this research aims to explore the opinions of Millennials on the use of live animals in events. The literature discusses anthropocentrism, anthropomorphism and cognitive dissonance, as reoccurring themes. A combination of a focus group and semi-structured interviews was undertaken, and the analysis identified entertainment, financial benefit and tradition as the main reasons for using live animals at events. Awareness and transparency on animal welfare issues within the events industry were stated by interviewees as points for improvement together with the lack of a clear definition of animal welfare, especially when it comes to captive and performing animals, as well as the uncertainty regarding animals' stakeholder status in events.

Introduction

Events are unique representations of culture and tradition and as such, they express and form people's attitudes and beliefs (Hall, 1997) and have the power to directly affect opinions and inspire change (Getz, 2005). Therefore, event organisers carry a certain amount of responsibility to reinforce positive social practices and behaviours and avoid those that are unethical and immoral (Bowdin, Allen, O'Toole, Harris, & McDonnell, 2011). Wilson (1984), suggests that people's psychological health is associated with their relationship to nature, a phenomenon called *biophilia*. Studies have confirmed this theory, showing that interactions with animals and feeling in harmony with nature offer health and well-being benefits to humans (Penn, 2003). This intrinsic desire to connect with the natural world can serve as an explanation for humanity's fascination with animals. Up until the early 1900s, the attitude towards animals was largely characterised by anthropocentrism, or the perceived superiority and exceptionalism of humans compared to the rest of the natural world (Garner, 1993). This view has been changing and evolving throughout the twentieth century and culminated in animal rights & welfare becoming a pivotal discussion in recent years. This shift in morals translates to lifestyle changes, such as identifying as a vegetarian/vegan and minimising one's consumption of products or services that include the use of animals. The number of people adopting a vegan lifestyle 'has doubled twice in the last 4 years' (The Vegan Society, 2018). This growth is believed to be a consequence of more information being publicly available about how animals are treated across different industries (Moss, 2016).

Both in theory and practice, stakeholder analysis plays a pivotal role in event management (Shone & Parry, 2010). Traditionally, a stakeholder is defined as any individual or organisation that has an interest in, or is influenced by, an organisation or project (Donaldson & Preston, 1995). Thus, anyone involved in the production, delivery and experience of an event is considered a stakeholder (Allen, O'Toole, Harris, & McDonnell, 2011). Therefore, it can be argued that when animals are involved in an event, they should be assigned a stakeholder status. Allen et al. (2011) state that an event's impact can be determined by looking at how effectively the needs of different stakeholders are met. This leads to the need to observe how, if at all, an animal's needs are identified and considered. One might argue therein lies the purpose of animal welfare legislation. However, the issue is that a universal definition of animal welfare does not exist (Haynes, 2008). According to Jasper and Nelkin (1992) animal welfare is not expressed in abandonment of using animals, but rather in ensuring less suffering is caused to them where they are used. An opposing perspective is 'animal liberation' – the belief that animals are entitled to moral consideration equal to that of humans, and capitalising on them should be discontinued (Haynes, 2008, p. 2). Dashper (2016, p. 23) argues the relationships between people and animals cross 'species, spatial, sensory and temporal boundaries' and goes on to explain these issues 'are complex and highly debated and no consensus has been reached amongst academics and practitioners'.

Getz (2012) states event research requires a multidisciplinary approach, studying culture, human behaviour, morals and sociology in order to be valuable. Furthermore, Jones (2014) states creating a lasting and sustainable event legacy is at the centre of producing events that nurture positive changes in society. Some aspects of ensuring sustainability, however, are less tangible and harder to measure, such as the effect on culture, communal thinking and consumer behaviour. Getz (2012, p. 91) argues people 'cannot be ethical or moral in isolation', thus, highlighting the impact one's social environment has on their moral philosophy. Getz proceeds to explore whether ethics is determined by law, or if morality brings an additional set of rules, beyond what is regulated by governing bodies. Despite this recognition of the importance for event management practices to be both sustainable and ethical, event theory discussing the ethics of animal use in planned events is limited. Many countries, including the United Kingdom, have introduced bans on performing animals in circuses, for example, yet other ways in which animals are involved in events remain permitted and widely unexplored. Considering the different beliefs about animal rights, this study aims to explore and understand Millennials' thoughts and feelings on using live animals in planned events. Millennials in the UK have been defined by Parliament as 'Roughly aged between 25 and 34' (Brown et al., 2017, p. 3) and they make up 13.9% of the total UK population. To achieve this, the following objectives were developed:

(1) To explore philosophies held by Millennials in relation to the natural world and animals in particular
(2) To discuss the reasons for animal use in planned events according to Millennials
(3) To observe participant's perceptions of animal treatment in the events industry
(4) To encourage participants to identify areas for improvement and ways to act on animal welfare issues at events.

Literature review

Nature-related philosophies

Debates on animal welfare date back to Antiquity. Ancient Greece offers varied opinions on the matter, the philosopher Pythagoras being the first known animal rights advocate. Pythagoras subscribed to animism – the belief that all components of the natural world (plants, humans, animals, land, etc.) are connected by a common spirit (Stringer, 1999). A similar philosophy is an ecocentrism which assigns a value to all living things and their habitats, unrelated to their usefulness to humanity (Gautam & Rajan, 2014). In contrast, another Ancient Greek philosopher, Aristotle, argued that human beings and animals share the same material features, but are distinguished by the mind (Adler, 1997). Aristotle believed only humans can think abstractly and demonstrate intelligence and thus, considered animals inferior to humanity (Adler, 1997). Later, during the French Renaissance, Montaigne argues animals are no less mentally capable than humans (Kenny, 2012), describing the complex behaviour of animals in support of his views. However, another French philosopher, Descartes, saw animals as mechanical beings, void of thought and feelings. This view is one of the components of anthropocentrism, which has filtered through history and is still adopted by many today (Steiner, 2005). Anthropocentrism is demonstrated by viewing humans as superior to other beings, by supporting the idea that humans exist separate from, and are not existentially connected to the rest of nature, and also by seeing the environment as a resource to be exploited (Xu & Fox, 2014).

Conversely, a philosophy which serves as a basis for much of today's arguments in favour of giving animals a moral status is ethical utilitarianism, made popular by the eighteenth-century English philosopher Jeremy Bentham. Bentham argues the righteousness of an action depends on whether it achieves the 'greatest happiness of the greatest number' (Bentham, 1789 cited in Schultz, 2017, p. 67). Bentham intentionally does not specify what the 'number' refers to, as he argues ethical consideration should be extended to other species based on the object's ability to experience suffering:

> ...the question is not, Can they reason? nor, Can they talk? but, Can they suffer? Why should the law refuse its protection to any sensitive being (Bentham, 1789 cited in Rollin, 2016, p. 11).

After more than 2500 years, there is still no consensus on what rights animals should be assigned (Ryder, 2000). Most of the work on the issue is written after the 1970s when the animal rights movement saw a dramatic increase in popularity (Adams, 2010). The catalyst for this is believed to be Richard Ryder's and later, Peter Singer's publications on the matter. In 1970 Ryder coined the term *speciesism* – putting one's own species' interests above those of other species – intentionally establishing a link with sexism and racism. Subsequently, Singer supported Ryder's views and argued speciesism is born from unjustified prejudice merely based on biological features, thus, suggesting that animals should be considered morally equal to humans. In their latest collaborative work, Ryder and Singer (2011) put forward the term of *painism* as a new ethical idea that stands for assigning moral rights to all living beings capable of suffering. Fennell (2012, p. 41) also explores the capacity for suffering as one of the variables that should be considered when debating animals' moral status whilst also discussing moral agency. Moral agency is the ability to assess the ethical implications of one's behaviour, which subsequently makes the agent responsible for the consequences of their actions. While Machan (2002) states human use of animals is justifiable due to animals' assumed lack of moral agency, Shapiro (2006) argues some non-human animals' observed behaviour when relating to other animals, both from their own and different species, serves as proof that non-human species can indeed demonstrate moral agency. Ethological research (the study of animal behaviour) shows animals manifest compassion, cooperation and deep emotions such as love and grief (Fennell, 2012). However, there has been very little research on this and the existing studies have not been given significant attention, possibly due to financial and professional interests (Fennell, 2012). Fennell (2012) further challenges human perception of animals' emotional abilities by suggesting that '*not being willing to understand the capacity for animals to feel is perhaps a limitation in the sensory capacity of humans.*' (Fennell, 2012, p. 43)

Many authors have tried to pinpoint the criteria upon which people judge the moral status and intrinsic value of animals. The factors prevalent in the existing literature are the ability to show empathy and emotion (Fennell, 2012; Singer, 2011), intellectual capabilities, autonomous thinking (Cochrane, 2009) and capacity for suffering (Bentham, 1789 cited in Fennell, 2012; Schultz, 2017; Singer, 1990).

Determinants of attitudes to animals

There are a number of studies (Galvin & Herzog, 1992; Curtin, 2006; Knight & Barnett, 2008; Apostol, Rebega, & Miclea, 2012) examining the causes of people adopting a certain view on animals. Gender is one determinant with women often manifesting greater concern for animal welfare issues (Apostol et al., 2012). Another factor is the ability to empathise with them (Apostol et al., 2012; Galvin & Herzog, 1992). Individuals demonstrating anthropomorphic views see animals as having qualities similar to humans and therefore find it easier to empathise with them. Thus, people with anthropomorphic tendencies approve of animal use less frequently than those who view animals as significantly different to humans (Galvin & Herzog, 1992). Galvin & Herzog investigated how a person's moral views affect their attitudes towards animal use through the Ethics Position Questionnaire (EPQ) developed by Forsyth (1980). Comparing animal rights activists to college students, Galvin and Herzog (1992) found ethical ideology also plays a significant part in how one relates to the natural world. Overall, animal rights activists demonstrate an 'absolutist' moral view of the world, which is characterised by high idealism and the belief ethical principles can be universally applied. The opposite view is relativism, the philosophy that whether an action is ethical or not, is to be judged on a case-by-case basis (O'Grady, 2002). As found by Galvin and Herzog (1992), college students not involved in animal rights campaigning expressed a more relativist philosophy in relation to animal welfare.

Knight and Barnett (2008) explored how people's views change depending on the animal's species as well as the purpose of use. Their findings show an individual's views are highly dependent on their experiences of interacting with animals – people were likely to oppose animal use if they considered the animal aesthetically attractive, more mentally and emotionally capable, or had spent time with an animal of the same species. This could be an explanation of people's admiration for some domesticated species in particular, such as dogs or cats, as they are more likely to have experience of them. Similarly, Daly and Morton (2009) found a correlation between spending time with animals and having anthropomorphic beliefs – a pet owner or someone who grew up with animals is more likely to perceive them as human-like. Participants in Knight and Barnett (2008) study gave the least approval to the use of animals for fashion, cosmetics, entertainment and sport. However, revisiting the overall findings from the same study suggests that the participants' views are influenced not only by the purpose of animal use but also by one's personal background. For example, some participants disapproved of fox hunting whilst being regularly involved in fishing. Higher levels of education on the topic and first-hand experiences have been linked to lower levels of support for many forms of animal use (Broida, Tingley, Kimball, & Miele, 1993; Pifer, Shimizu, & Pifer, 1994; Knight & Barnett, 2008). Even though there is a significant lack of knowledge and research on animal ethics, making information available and educating the public does not come without challenges. Knight and Barnett (2008) found people deliberately avoid upsetting information about animal use, realising such knowledge might prevent them from enjoying elements of their daily life. Behaviour such as this was described as cognitive dissonance by Festinger (1957).

In a study by Curtin (2006) similar findings are demonstrated after investigating people's experiences of swimming with dolphins and the changes in their view of the species because of the encounters. The research involved interviewing people who swam with captive dolphins (SCD) in a controlled environment, and others who had an encounter with the animals in the wild (SWD). She found that witnessing animals in their natural habitat is more satisfying to humans than having watched captive animals. Comparing SWD respondents to the SCD group reveals participants' different views on the way humans relate to nature. SWD participants appear less anthropocentric, considering captivity unethical. The SCD group express similar concerns, but they attempt to justify it by focusing on the animals' emotional connection with their trainers and the care dolphins receive. In other words, SCD interviewees demonstrated cognitive dissonance expressed in their avoidance of information that can add to the uncomfortable feelings associated with contributing to captivity (Festinger, 1957). Curtin concludes that the dissonance resulting from encounters with captive dolphins is a result of the participants' anthropomorphic image of dolphins, expressed in the perception animals derive comfort from their relationship with their trainers. Her analysis creates a different perspective on anthropomorphism, directly opposing the findings discussed earlier, where anthropomorphism was found to have a positive effect on reducing animal use due to feelings of empathy (Galvin & Herzog, 1992).

Animals in culture, tradition & events

Leventi-Perez (2011) argues human perception of animals is shaped by external factors – the anthropomorphic representation of animals in Disney films, for example, which is characterised by attributing human qualities and behaviours to unhuman beings, e.g. animals & plants (Serpell, 1996). Anthropomorphic images are arguably ingrained in the human mind as a result of animal portrayal in popular media. This phenomenon, discussed by De Waal (2001, p. 71), is described as 'bambification' – stripping wildlife of primal instincts, generally perceived as negative, and building an animal's image around 'cuteness' and other more marketable qualities. Arguably, as observed earlier from Curtin's (2006) and Knight and Barnett (2008) studies, such representation of animals is double-edged – while it might help viewers relate to animals, it does little for the connection between humans and 'real' nature. As a result, people's understanding of true animal behaviour is distorted which leads to further detachment from the natural world (Leventi-Perez, 2011). Fennell (2017) discusses the role of animals in ecotourism, highlighting the issues stemming from their representation in tourism advertisements. Promotion images aim first and foremost to appeal to tourists, whilst looking authentic to local culture, thus, neglecting the consequences of the portrayal of animals as 'passive and secondary to the tourism experience' (Fennell, 2017, p. 186).

Planned events are both the product and expression of tradition, communal values and identity (Liutikas, 2016). Spracklen and Lamond (2016) argue that events, especially when coupled with media, play a vital role in the spreading, normalisation and perpetuation of ideologies & values. Therefore, similarly to media, events can have a significant effect on the way people see and treat animals.

The use of performing wild animals in circuses and marine parks has been banned in the UK and many other European countries. However, it remains a practice overseas under the guise of education (Donaldson & Kymlicka, 2011). Sugarman (2007) states that pressure from animal rights groups threatens circuses' legacy, resulting in economic difficulties for establishments using animals. However, Sugarman fails to address why such a legacy is important, also, what is the quality of life of the animals themselves; their welfare beyond numbers of years lived. Donaldson and Kymlicka (2011) argue that circuses, zoos and marine parks are indeed involved in education but the lessons taught are not love for, and knowledge of the natural world, but rather disrespect to animals' freedom and promotion of human entitlement and superiority. According to Jaynes (2008) circuses with performing animals are loved and attended by many due to their nostalgic value originating from the attendees' childhood memories. This sentiment leads Jaynes (2008, p. 5) to question, 'why do we think we have the right to force animals into these situations?' This study, therefore, explores the philosophies held by Millennials in Britain aiming specifically in relation to animal treatment in the events industry.

Methodology

As the study is exploratory, drawing on theory from different disciplines, a qualitative approach was selected. Two data collection methods were used with nine participants in a focus group and three in-depth interviews. The focus group method was selected due to the relative novelty of the topic and by encouraging a group discussion it would offer the opportunity for different opinions and arguments to emerge and develop in a dynamic conversation, closely mimicking a natural discussion (Krueger & Casey, 2000). Building on the work of Dashper (2016) on human–horse relationships, statistical information about the British horse racing industry (see Table 1) was read to participants in the focus group, providing a starting point for the discussion and drawing on the literature, questions were developed for both the focus group and interviews (also see Table 1). The focus group helped determine general attitudes and patterns towards the use of animals in events, which were then further investigated through conducting semi-structured interviews. The interviews followed the same structure and questions as the focus group discussion – participants were first asked about their philosophies relating to nature and animals. Thereafter, the researcher directed the conversation towards the interviewee's view of animal use specifically at events. Specific insights from the focus group which were chosen for further discussion in the interviews were whether animals should be assigned an event stakeholder status and also what is the definition of animal welfare/fair treatment.

Millennials were chosen as the study sample for both data collection methods, as more people aged 15–34 than any other generation, have been found to be concerned with the human impact on animals and the environment (Perlis, 2016). Adopting the sampling approach of Hill, Mobly, and McKim (2016), students, aged 18 and over, resident to southern England were identified as an opportunistic sample and were recruited by contacting student groups on social media. To achieve a heterogeneous sample, socio-demographic variables such as age, gender, country of origin, course studied and lifestyle (vegan/vegetarian/omnivore) were considered. Approximately 2% of the UK population are vegetarian or vegan (NHS, 2015), with almost half of those being Millennials (The Vegan Society, 2017). Therefore, a small number of vegetarian/vegan participants were included in the focus group, reflecting the proportions of vegetarians/omnivores in British society.

The focus group and semi-structured in-depth interviews were undertaken in the spring of 2017, audio recorded and transcribed. All the data were combined and analysed thematically (Fox, Gouthro, Morakabati, & Brackstone, 2014) to identify patterns in responses relating to existing theory, as well as any new insights. Ethical approval from the researchers' institution was obtained prior to the data collection. This included guaranteeing the participants' anonymity throughout and therefore numbers have been assigned to each participant in the next section (P1–P9 from the focus group and P10–12 for the interviewees).

Table 1. British horse racing statistics and sample questions.

The British horse racing industry is currently worth around £3.45 billion annually (British Horse Racing Authority, 2017).
Racehorses are bred in a way that makes them genetically faster but also susceptible to health problems as a result of their unnaturally thin bones (BBC, 2007).
Two thirds of horses bred for racing never even enter the industry. Furthermore, only around 100 of the approximately 5,000 horses retiring from racing each year are taken into care. In 2010, more than 7,000 British horses were slaughtered and sold for consumption to other European countries, primarily Belgium and France, a number that has been steadily rising in the past two decades (BBC, 2007).
The Royal Society for the Prevention of Cruelty to Animals (RSPCA) states that 'using whips can cause pain and suffering to the horses' but whipping the horses the entire duration of a race is a standard practice, which is banned in Norway, for example (Clark, 2014).

Sample questions		
Subtopic	General views on the relationship between humans and the environment	Perceptions of the treatment of animals in events	Animal use in traditional events: e.g. British Horse Racing & Bull Fighting in Southern Europe
Example question	*Do you think we should take care of animals and the environment for their own sake, or because of the value they provide to us?*	*What are the moral considerations that should be kept in mind when using animals in events?*	*What is the horse's role in horse racing? Are they there because they enjoy racing or because they are forced to do it?*

Findings

Both the focus group and interviews began by asking participants about their views concerning nature and animals. This helped to set the background and create a picture of how British Millennials relate to, and think, of nature. Generally, participants demonstrated a strong concern for the protection of the natural world and animals. Some of the participants mentioned adopting a vegan/vegetarian lifestyle as their way to reduce their negative impact on the environment and animals. Even though all participants agreed that looking after the environment is important, they differed in their reasoning and motivation. Most believe that '*the environment and our planet can live without us but we can't live without the planet and the animals*' [P3], expressing an opinion that humans '*exist in a symbiotic relationship*' [P11] with the rest of nature. Some participants demonstrated a particularly eco-centric philosophy, describing a sense of connectedness and belonging to nature, for example:

> We come from nature so it's a matter of respect as well... I feel like if you respect what was given to you, you are more likely to be a better person and live in a better world, because you are more compassionate towards what's around you. [P6]

Participants also stated they do not agree with '*using animals for our own benefit*' [P10], directly opposing anthropocentric views: '*...we're not a supreme species. We're not the ones who own the planet, we share it, so, we should share it equally.*' [P2]. These responses, therefore, express strong biocentric opinions, believing the environment and other creatures have value in themselves, regardless of their usefulness to humans. Most participants assigned moral value to animals based on their capacity for experiencing pleasure and pain (supporting Singer, 2011, Fennell, 2012):

> Animals are creatures as well, they feel too. You like living a good life, don't you? It's basically the same. [P12]

When discussing the reasons for animal use in planned events, novelty, '*mass entertainment*' [P11] and the '*pure enjoyment of watching the animals do tricks*' [P10] were stated. Participants suggested that people attend such events to satisfy their curiosity and see species they would not normally encounter: '*I think it's the new experience that they [people] find entertaining*' [P10]. Moreover, performing animals are perceived as more relatable and '*humanised*' [P10]. This belief that animals possess human characteristics or anthropomorphism is related to showing empathy towards animals (Knight & Barnett, 2008) and was observed in this research: '*animals have so many human qualities... they have emotions.*'[P1]. Knight & Barnett's results also showed that people are more likely to empathise with animals and be concerned about animal welfare if they have had experiences with them. However, modern lifestyles, especially in Western countries, have made humans largely disconnected from the natural world. The Internet, particularly videos shared on social media, was identified by participants as their main source of information when it comes to animal welfare. This is not surprising given that online sharing has been found to be the main and preferred way of communication for Millennials (Pew Research Centre, 2010).

Anthropomorphic perceptions, resulting from animal portrayal in mass media, marketing and even wildlife documentaries may lead to people having certain unrealistic expectations from their encounters with animals which is reflected in the study by Curtin (2006). One interviewee's account illustrates this:

> If you have ever been on a safari, for example, what do the animals do there? They are just relaxing. People don't want to see that, they want to see animals chasing other animals, doing tricks and such, so maybe that's why people prefer attending shows. [P12]

Consistent with previous research (Galvin & Herzog, 1992; Curtin, 2006 and Knight & Barnett, 2009), one's personal background and experiences with animals is found to significantly affect their views. This was confirmed by a participant's narrative about their upbringing and background in relation to horse racing:

> ... then I started thinking, maybe I have been brought up with this so I'm biased and I think it's ok as we take really good care of the horses, but at the same time, why are we on top of the horse, why do we need to do that? [P6]

Both in the focus group and interviews, the economic aspect of using animals emerged as a central issue. Whilst some participants considered capitalising on animals unfair and immoral, others saw animal-related events, such as horse racing, as providers of valuable employment. The participants expressing views in support of animal use considered a balance can be achieved between animal welfare and financial profit:

> I think it's something that's been around for such a long time in Britain, it just brings so much money in... and employs hundreds of thousands of people. So, I think we can find ways of treating them better [the horses], without it affecting [the economy] ... There are ways to make it nicer for the horses without it affecting people's jobs. [P3]

The other group demonstrated views characterised by high moral absolutism and idealism, stating that using animals for profit is morally wrong as they *'don't belong to us'* [P6]. Thus, those participants who considered animal use as essential to a healthy economy, recognised the financial benefits to compromising animal welfare. Even though the focus group began with participants expressing strong absolutist views in support of environmental protection and animal rights, as discussions progressed, more relativist statements could be observed, that is that moral values and principles are not applicable to every situation (O'Grady, 2002). For example, at the outset, all participants identified themselves as aspiring to live in harmony with the natural world, but only a small proportion then stated they had taken specific action in that direction.

The significance of culture and tradition to animal use in events was examined by encouraging participants to compare horse racing events in Britain to bullfighting in Southern Europe. Bullfighting evoked very strong responses:

> ... bullfighting is supposed to be a tradition, but we're in 2017 and animals should not be treated that way. Those bullfights are barbaric and there's not much of an excuse to treat animals like that. [P3]

On the other hand, horse racing was subject to a more detailed conversation that produced a wider variety of views. All participants consider horse racing as more ethical with some stating there is no room for comparison with bullfighting:

> ... people have been riding horses for ages! ... not all horses are beaten, they do enjoy being trained and ridden. [P1]

Statistical information highlighting welfare issues in the horse racing industry was presented to participants, including fatal injuries. When comparing bullfighting and horse racing, if either form of event had fatal results for the animal, participants considered horse racing more ethical and morally acceptable as the initial intention was not to hurt the animal, whereas in bullfighting, killing the bull is the end goal: *'The focus is not on hurting the horse'* [P6]; *'It's not a violent thing [horse racing]'* [P5]. Even though concern was expressed over the fate of horses in case of an injury resulting in inability to race, some participants stated that there are significant economic barriers to discontinuing horse racing events. Interestingly, these were not mentioned as an issue when discussing the ending of bull-fighting, which was referred to as primitive and aggressive. However, culture and tradition can significantly affect one's views and most participants had a British background and therefore, the cultural environment might be a factor influencing the opinions expressed.

A more general negativity in response to animal treatment in events was also observed. Participants generally stated that using animals for entertainment is not justified, keeping in mind some of the training and conditioning methods involved:

... you see these caring aspects [of the animals] that we can relate to, and then they are being whipped for entertainment.... You wouldn't do that to people, why should you do that to somebody else when they clearly care for one another ... [P5]

However, most of the participants also believed if the animals are not mistreated, using them is not immoral. When questioned how one determines if an animal is well looked after, most participants expressed opinions that can be summarised as 'being treated well and not being forced to do anything' [P7] as criteria for animal welfare. Yet, an issue prevalent in the group discussion was the very nature of using animals, which by default involves exerting some amount of control over the creature and limiting their will. This was echoed in an interview:

you can't really tell with animals [if they are happy] ... they are being made to do tricks, whether they like it or not. [P12]

An emerging issue is the regimented nature of events involving animals – shows and races take place at a certain time, requiring animals to perform on demand. Thus, training and conditioning is needed, much like the preparation, people need in order to take part in a competition or do a certain job. However, animals are not being recognised or rewarded for their efforts, at least not in the same way people are (Singer, 1990). Some participants who had been involved in animal-related events, such as horse racing, mentioned the training and conditioning required to get the animal to respond to human command:

We used to train horses since they were young and they would do whatever we want them to do. In the beginning, they wouldn't, but that's when you start using the whips. [P6]

Conditioning as a major part of training animals in sport and events was discussed by participants with disapproval:

I guess people always say that they [the horses] enjoy it and think they're having fun but they're being whipped, it's not as extreme as bull fighting but there is cruelty involved. [P2]

Additionally, the notion of important differences between how animals are portrayed in an event and how they are treated 'behind the scenes' was particularly common in the discussion.

Significant differences could also be observed within the group when participants were asked to reflect on events involving animals kept as pets, such as dog shows. Treatment of domesticated animals is seen as fairer and more of a symbiotic relationship beneficial for both parties, where animals provide companionship to humans in exchange for shelter and care: 'I guess people give animals life in exchange for company' [P4]. In contrast, when reflecting on the use of wild animals in shows, a popular belief among participants was that humans cannot replicate an animal's natural habitat: 'animals always look happier in the wild' [P8]. A further problem is posed by the fact people 'haven't reached that point in our development when we are able to communicate with other species' [P11]. Therefore, participants who were strongly in favour of animal rights, raised the issue of the present ambiguity around compliance:

None of these animals can give consent... humans, it seems, we take advantage, and that's why it's such a big question, can we take advantage? We are just another species, we are not that much superior. [P2]

Participants often explained their diminishing support and opposition to animal use in planned events, as a result of gaining more information on the ethical issues involved in such practices. Thus, lack of transparency and limited knowledge about animal welfare issues in the event industry were highlighted.

Whether animals are considered stakeholders is unclear, an issue which further revealed itself in the study. The stakeholder status of animals needs to be evaluated if people are to ensure their welfare. Interestingly, discussing horse racing as an animal-related event came as a surprise to some participants with one interviewee stating that they 'have never actually thought of horse racing as an animal-related event.' [P12]. Another participant's account gives further insight into the problem:

't's really not about the horses, it's about going out, dressing up, betting, having a drink, socialising. And I feel like the horses are just there, but normally people wouldn't have any interest in horses....They're just a means to an end for making money.... You see so many horses dying in these events, seen as expendable. If a horse dies, they replace it. I really don't agree with that. [P10]

The participants considered that it is the event organisers' responsibility to provide transparency and information on animal welfare issues. However, giving more value to financial motivators rather than animal welfare is identified as a barrier for companies in acting on this responsibility and spreading awareness:

... it's the commercial organisations' [duty to inform], because the animals are in their care – if people put pressure on them for that transparency, they may get it, but if the organisations make more money from lack of transparency, they wouldn't sacrifice their profit for the sake of the animals.... That's what businesses are about today, just making more money than they did yesterday. [P11]

Discussion, conclusion & recommendations

The aim of the research was met through the exploration of a wide range of topics relating to the use of animals in events, by British Millennials. This included their opinions on the natural world, their perceptions of animal treatment within the events industry and their views on how current animal welfare issues can be resolved. The complexity of the topic became apparent as it entailed debating animal welfare and animal rights issues, which are sensitive topics, demanding in-depth discussion.

Most participants expressed opinions that can be described as biocentric. However, a wide range of nuances in people's philosophies was revealed through observing the participants' reasoning behind holding a certain view concerning animal involvement in events. People's innate fascination with, and curiosity about nature are some of the reasons to use animals in ways, not critical to one's survival, such as planned events. Tradition, such as horse racing, and cultural festivities, such as bullfighting, are other motivators.

Some key conflicts were identified while looking at Millennial's philosophies regarding the natural world and their opinion on animal use in events. One issue is the effects of anthropomorphism. Even though anthropomorphic tendencies are related to a greater capacity for empathising with animals and opposing their exploitation, a contradiction stems from the fact these views seem to be related to engaging in activities involving animals. Some people develop greater empathy to animals and consider their use unfair, only after witnessing first-hand what is involved in having an animal at an event. However, adopting an anthropomorphic view of animals can lead to an increased demand for animal-related events, due to people's limited knowledge of animals' true nature, consisting of instincts and needs humans cannot always anticipate.

Another problem in need of attention is the lack of clear definition of animal welfare, which currently, seems to be dependent on individual interpretation. Animal rights groups and individuals with absolutist moral views advocate for 'animal liberation', condemning all animal use across different industries (Haynes, 2008), whereas others adopt a more pragmatic approach, claiming reducing animal suffering is enough to ensure welfare (Jasper & Nelkin, 1992). A question that emerged from participants' responses is whether people are entitled to use animals for entertainment purposes, whilst being unsure of how well their needs are met.

Economic factors seem to be central to perpetuating animal use at events. Participants stated demand could be lessened by raising awareness of animal welfare issues and educate the public about the natural world, encouraging more ethical and sustainable practices within the event industry. However, previous research reveals cognitive dissonance associated with the use of animals in entertainment causes people to avoid upsetting information on the issue. As time progresses, people are more informed and more open to assigning rights to other species, which is evident in the progression of opinions expressed in more recent studies and literature, compared to those from a few decades ago.

'How can we care about species we have never seen?' is a question often used to challenge anti-captivity campaigners and animal rights activists (Russo, 2013). However, the adoption of an ecocentric perspective, which was found to be popular among Millennials, makes it of no importance whether one has seen a certain species of animal or not, whether one cares about that animal or not, what matters is the responsibility to allow other species to pursue life and survival on their own terms (Bekoff, 2013). Humans are neither destructive villains, nor protectors of the natural world, but rather an integral part of it (Smythe, 2014).

Limitations of the study included the relatively small number of participants and the nature of discussing sensitive morality-related topics. When talking in groups about ethics & morals participants might feel obliged to conform to a certain 'ideal' that is considered morally 'right' or more widely spread than other views (Barbour, 2007). As the research topic touches on the participants' moral & ethical values concerning the treatment of animals at events, responses gathered through the focus group discussion might not reflect views in their entirety due to concerns of how one is perceived by other participants.

A question for further research is how to bridge the gap between the ideology people express and consumer behaviour – the most prevalent views demonstrated were those related to ecocentrism and the claim that animals need to be recognised as deserving fair treatment. However, as observed by Auger and Devinney (2007), these beliefs are not always brought into practice. As there is a recognised need for more education on animal welfare within events, challenges to raising awareness, such as cognitive dissonance and avoidance of distressing information, should be explored in further research. Additionally, a comparison could be made with participants from different cultures, such as young Spaniards who have grown up within a culture of bullfighting. The findings from this exploratory study could also be developed further through a quantitative study which measures the attitudes and may lead to a segmentation of Millennials.

In conclusion, despite event legacy having gained a central role in recent event studies (Jones, 2014), there is little theory covering the moral aspects of the human-animal relationship. Event theory places great emphasis on the importance of stakeholders (Allen et al., 2011; Getz, 2005), yet, animal use in events is rarely mentioned in event or leisure literature. Findings suggest animals are not currently considered event stakeholders, which requires further exploration. Finally, policymakers need to address the lack of a clear definition of animal welfare, which, at present seems surrounded by ambiguity and dependant on personal philosophy.

Disclosure statement

No potential conflict of interest was reported by the authors.

References

Adams, C. (2010). The war on compassion. *Antennae, 14*, 5–11.

Adler, M. J. (1997). *Aristotle for everybody: Difficult thought made easy*. New York: Touchstone Books.

Allen, J., O'Toole, W., Harris, R., & McDonnell, I. (2011). *Festival & special event management* (5th ed.). Milton: John Wiley & Sons Australia.

Apostol, L., Rebega, O. L., & Miclea, M. (2012). Psychological and socio-demographic predictors of attitudes toward animals. *Procedia - Social and Behavioral Sciences, 78*, 521–525.

Auger, P., & Devinney, T. M. (2007). Do what consumers say matter? The misalignment of preferences with unconstrained ethical intentions. *Journal of Business Ethics, 76*, 361-383. doi:10.1007/s10551-006-9287-y

Barbour, R. (2007). *Doing focus groups*. Los Angeles: Sage Publications.

BBC. (2007). Ex-race horses should be slaughtered for meat. Retrieved from http://www.bbc.co.uk/somerset/content/articles/2007/12/04/horseworld_feature.shtml

Bekoff, M. (2013). Animal consciousness and science matter: Anthropomorphism is not anti-science. *Relations beyond Anthropocentrism, 1*(1), 1–8.

Bowdin, G., Allen, J., O'Toole, W., Harris, R., & McDonnell, I. (2011). *Event management* (3rd ed.). Oxford: Elsevier Butterworth-Heinemann.

British Horse Racing Authority. (2017). *About us*. British Horse Racing Authority. Retrieved from http://www.britishhorseracing.com/bha/about-us/

Broida, J. P., Tingley, L., Kimball, R., & Miele, J. (1993). Personality differences between pro and antivivisectionists. *Society and Animals, 1*, 129–144.

Brown, J., Apostolova, V., Barton, C., Bolton, P., Dempsey, N., Harai, D., … Powell, A. (2017). *Millennials*. Retrieved from www.parliament.uk/commons-library

Clark, A. (2014). *The grand national: 8 things they don't tell you about horse racing*. PETA UK. Retrieved from http://www.peta.org.uk/blog/the-grand-national-8-things-they-dont-tell-you-about-horse-racing/

Cochrane, A. (2009). Do animals have an interest in liberty? *Political Studies, 57*, 660–679.

Curtin, S. (2006). Swimming with dolphins: A phenomenological exploration of tourist recollections. *International Journal of Tourism Research, 8*, 301–315.

Daly, B., & Morton, L. (2009). Empathic differences in adults as a function of childhood and adult pet ownership and pet type. *Anthrozoös, 22*(4), 371–382.

Dashper, K. (2016). *Human-Animal relationships in equestrian sport and leisure*. London: Routledge.

De Waal, F. B. M. (2001). *The ape and the Sushi master: Cultural reflections of a primatologist*. New York: Basic Books.

Donaldson, S., & Kymlicka, W. (2011). *Zoopolis: A political theory of animal rights*. Oxford, UK: Oxford University Press.

Donaldson, T., & Preston, L. (1995). The stakeholder theory of the corporation: Concepts, evidence, and implications. *Academy of Management Review, 20*(1), 65–91.

Fennell, D. A. (2012). *Tourism and animal ethics*. New York, NY: Routledge.

Fennell, D. A. (2017). *Tourism Ethics*. Bristol: Channel View Publications.

Festinger, L. (1957). *A theory of cognitive dissonance*. Evanston, IL: Peterson.

Forsyth, D. R. (1980). Taxonomy of ethical ideologies. *Journal of Personality and Social Psychology, 39*(1), 175–184.

Fox, D., Gouthro, M. B., Morakabati, Y., & Brackstone, J. (2014). *Doing events research: From theory to practice*. London: Routledge.

Galvin, S., & Herzog, H. (1992). Ethical ideology, animal rights activism, and attitudes toward the treatment of animals. *Ethics & Behaviour, 2*(3), 141–149.

Garner, R. (1993). *Animals, politics and morality*. Manchester: Manchester University Press.

Gautam, C. K., & Rajan, A. P. (2014). Ecocentrism in India: An incredible model of peaceful relation with nature. *Universal Journal of Environmental Research & Technology, 4*(2), 90–99.

Getz, D. (2005). *Event management and event tourism* (2nd ed.). New York: Cognizant Communications Corporation.

Getz, D. (2012). *Event studies; Theory, research and policy for planned events*. Abingdon: Routledge.

Hall, C. M. (1997). *Hallmark tourist events: Impacts, management and planning*. Chichester: John Wiley and Sons.

Haynes, R. (2008). *Animal welfare: Competing conceptions and their ethical implications*. London: Springer.

Hill, J., Mobly, M., & McKim, B. (2016). Reaching millennials: Implications for advertisers of competitive sporting events that use animals. *Journal of Applied Communications, 100*(2), 73–85.

Jasper, J. M., & Nelkin, D. (1992). *The animal rights crusade: The growth of a moral protest*. New York: Maxwell Macmillan International.

Jaynes, M. (2008). The ethical disconnect of the circus: Humanity's acceptance of performing elephants. *Between the Species, 8*, 1–11.

Jones, M. (2014). *Sustainable event management: A practical guide*. London: Routledge.

Kenny, A. (2012). *A new history of Western philosophy*. Oxford: Oxford University Press.

Knight, S., & Barnett, L. (2008). Justifying attitudes toward animal use: A qualitative study of people's views and beliefs. *Anthrozoös, 21*(1), 31–42.

Krueger, R. A., & Casey, M. A. (2000). *Focus Groups: A practical guide for applied research* (3rd ed.). London: Sage Publications.

Leventi-Perez, O. (2011). *Disney's portrayal of nonhuman animals in animated films between 2000 and 2010* (Master's thesis). Scholar Works@ Georgia State University, Atlanta, USA.

Liutikas, D. (2016). Indulgence feasts: Manifestation of religious and communal identity. In A. Jepson & A. Clarke (Eds.), *Managing and developing communities, festivals & events* (pp. 148–149). Basingstoke: Palgrave McMillan.

Machan, T. R. (2002). Why human beings may use animals. *Journal of Value Inquiry, 36*(1), 9–16.

Moss, R. (2016). *Number of vegans in Britain soars in past decade, here's why*. The Huffington Post UK. Retrieved from http://www.huffingtonpost.co.uk/entry/number-of-vegans-in-uk-half-million_uk_573c2557e4b0328a838b92a3

NHS. (2015). Healthy eating for vegetarians and vegans. Retrieved from: http://www.nhs.uk/Livewell/Vegetarianhealth/Pages/Goingvegetarian.aspx

O'Grady, P. (2002). *Central problems in philosophy: Relativism*. London: Routledge.

Penn, D. J. (2003). The evolutionary roots of our environmental problems: Toward a Darwinian ecology. *The Quarterly Review of Biology, 78*(3), 275–301.

Perlis, M. (2016). Forbes survey reveals what millennials really want. Retrieved from https://www.forbes.com/sites/mikeperlis/2016/06/06/forbes-survey-reveals-what-millennials-really-want/#60d186a36755

Pew Research Centre. (2010). Millennials: A portrait of generation next. Confident. Connected. Open to change. Retrieved from http://www.pewsocialtrends.org/files/2010/10/millennials-confident-connected-open-to-change.pdf

Pifer, L., Shimizu, K., & Pifer, R. (1994). Public attitudes towards animal research: Some international comparisons. *Society and Animals, 2*(2), 95–113.

Rollin, B. E. (2016). *A new basis for animal ethics: Telos and common sense*. Columbia: University of Missouri Press.

Russo, C. (2013). *Can you worry about an animal you've never seen? The role of the zoo in education and conservation*. Sci-Ed. Retrieved from http://blogs.plos.org/scied/2013/03/11/zoo-education/

Ryder, R. D. (1970). Speciesism. Retrieved from http://www.criticalsocietyjournal.org.uk/Archives_files/1.SpeciesismAgain.pdf

Ryder, R. D. (2000). *Animal revolution: Changing attitudes towards speciesism*. London: Bloomsbury.

Ryder, R. D., & Singer, P. (2011). *Speciesism, painism and happiness: A morality for the twenty-first century*. Exeter: Imprint Academic.

Schultz, B. (2017). *The happiness philosophers: The lives and works of the great utilitarians*. Princeton, NJ: Princeton University Press.

Serpell, J. (1996). *In the company of animals: A study of human-animal relationships*. Cambridge: Cambridge University Press.

Shapiro, P. (2006). Moral agency in other animals. *Theoretical Medicine and Bioethics, 27*(4), 357–373.

Shone, A., & Parry, B. (2010). *Successful event management: A practical handbook*. Andover: Cengage Learning.

Singer, P. (1990). *Animal liberation* (2nd ed.). London: Jonathan Cape.

Singer, P. (2011). *Practical ethics* (3rd ed.). Cambridge: Cambridge University Press.

Smythe, K. (2014). Rethinking humanity and the Anthropocene: The long view of humans & nature. *Sustainability: the Journal of Record, 7*(3), 145–153.

Spracklen, K., & Lamond, I. R. (2016). *Critical event studies*. Abingdon: Routledge.

Steiner, G. (2005). *Anthropocentrism and its discontents: The moral status of animals in the history of Western philosophy*. Pittsburgh, PA: University of Pittsburgh Press.

Stringer, M. D. (1999). Rethinking animism: Thoughts from the infancy of our discipline. *Journal of the Royal Anthropological Institute, 5*(4), 541–556.

Sugarman, R. (2007). *The many worlds of circus*. Newcastle: Cambridge Scholars Publishing.

The Vegan Society. (2017). Key facts. Retrieved from https://www.vegansociety.com/about-us/key-facts

The Vegan Society. (2018). *Key facts*. The Vegan Society. Retrieved from https://www.vegansociety.com/about-us/key-facts

Wilson, E. O. (1984). *Biophilia*. London: Harvard University Press.

Xu, F., & Fox, D. (2014). Modelling attitudes to nature, tourism and sustainable development in national parks: A survey of visitors in China and the UK. *Tourism Management, 45*, 142–158.

A predator in the park: mixed methods analysis of user preference for coyotes in urban parks

Jackson Wilson (iD) and Jeff Rose (iD)

ABSTRACT

This mixed methods study explores the relative preference people in the United States have for sharing leisure space in their local urban parks with coyotes. Two rounds of survey data ($n= 482$) and a series of interviews ($n= 28$) were conducted. In both survey samples, people preferred to share park space with coyotes less than all other species options (e.g., people experiencing homelessness, off-leashdogs). Interview data suggest that the primary reason for this lack of desire to share park space with coyotes is a perception that coyotes are dangerous for people and pets. This strong level of preference against coyotes has implications for current efforts to promote human-wildlife coexistence strategies in many urban and peri-urban locations.

Parks are not only physical, material spaces of leisure; they are also social constructs (Proctor, 1998). Social norms dictate appropriate behaviours in parks and subsequently constrain which humans and other species are welcome or spurned from these natural urban and peri-urban spaces (Taylor, 1999; Webster, 2007). These norms are often contentious and create public debates about whether different visitors, including off-leash domestic dogs (Wilson, Yoshino, & Latkova, 2018) or people experiencing homelessness (Rose, 2017), should have the same access to public parks as housed people that will abide by middle-class standards of behavior (e.g., relatively clean clothes, innocuous smell, visiting the park for supposedly *appropriate* recreational purposes; Taylor, 1999).

The coyote (*Canis latrans*) has trotted into the middle of this debate (Figure 1). Coyotes are a North American canid that are 1–1.5 m in length and 9.1–16.0 kg in weight (Bekoff & Gese, 2003). Aided by anthropogenic changes to the environment, such as the elimination from wolves from much of their traditional range, coyotes have expanded their range from Northern Alaska to Costa Rica (Bekoff & Gese, 2003; Rutherford, 2018). That increase in range includes urban and peri-urban areas that have traditionally been socially constructed as spaces dedicated for people (Rutherford, 2018; Wolch, 2002).

The presence of coyotes in urban environments has led to a range of responses. Although rare, coyotes occasionally attack humans (Baker & Timm, 2017). In contrast, they regularly prey upon domesticated cats and small dogs (Alexander & Quinn, 2011). Fear of coyotes has classically resulted in attempts to exterminate them (Timm & Baker, 2007). More recently, agencies have adopted less lethal management stances (Alexander & Quinn, 2011; White & Gehrt, 2009). Rather than trapping and killing coyotes, the goal has been to educate the public on how to minimize negative interactions. However, there are some that believe this education-based management strategy is not appropriate for parks and other urban green spaces and coyotes should be removed from all urban and peri-urban locations (Webster, 2007).

Figure 1. Galante (2018) urban coyote.

There is conflicting evidence about current North American attitudes towards coyotes in local parks. In a US-based nation-wide survey, Kellert (1985) found that dogs were the most liked and coyotes were one of the less liked animals, but he also commented that the 'traditional European antipathy toward the predator' (p. 168) was waning, and Americans were moving towards a 'far greater diversity and conflict of viewpoints, paralleling America's divisions on wildlife, or the necessity of wildlife damage control' (p. 168). Baker and Timm (1998) described this shift as, '... attitudes of many people in today's society toward wild animals have changed from respect and fear to a certain reverence' (p. 309). Timm (2006) characterized this shift as the public 'developing (a) "Disney mentality" toward wildlife' (p. 144). Although these statements suggest a growing tolerance of coyotes, an increasing number of attacks (Baker & Timm, 2017) and negative evaluations of coyotes in the media (Alexander & Quinn, 2012) may reverse this trend. As Gehrt (2004) equivocally stated, 'Although the sight or sound of a coyote elicits fear in some residents, it often produces more favorable responses in others' (p. 84).

Given the lack of clarity about the public's perception of coyotes in their local parks, this study examined the relative preference for sharing parks and other urban green spaces with coyotes using data from a survey that compared the preference for sharing park space with other entities and interviews to further explore those rankings. These analyses inform contemporary perspectives of wildlife in urban space and provide data to inform park management practices. This research also helps characterize human-animal relations in settings that are most typically characterized as leisure spaces.

Methods

A sequential mixed method design was used, grounded in pragmatism (Creswell & Creswell, 2018), a paradigm that focuses on consequences. Operating from this perspective often leads researchers to use multiple methods to investigate the research question. This study used surveys to understand preferences, and conducted interviews to help explore those rankings. Survey samples ($n= 482$) were collected online and in-person using an instrument developed for this study to understand the relative preference of people to share park space with different entities. A series of interviews ($n= 28$) further explored participant preferences for sharing public park spaces with coyotes.

Instrument

The instrument was developed to understand the relative preference of people for sharing public park space with eight different entities; person, person perceived as experiencing homelessness, person with on-leash dog, person with off-leash dog, homeless person with on-leash dog, homeless person with off-leash dog, off-leash dog with no person, and a coyote. The eight entities were chosen based on current controversies focused on the presence of dogs (Wilson et al., 2018), homelessness (Rose, 2017), and coyotes (Webster, 2007) in urban and peri-urban parks.

The instrument had five sections. After the informed consent, respondents were asked, 'What is the name of the park that you most frequently visit for 15 minutes or more?' and then asked to identify the location of that park. The name of each respondent's primary park was inserted into subsequent questions about the park to provide respondents with a concrete, known context to ground their responses. Second, respondents were presented with two attention-checking dichotomous survey items. These two items described one of two pictures and requested the respondent to click on the described picture. The third section started with an introduction informing the respondent that there would be a series of 28 unique items asking which entity they would prefer to see in their primary park they had previously identified. The 28 dichotomous items forced respondents to choose the most preferred of each of the eight entities (e.g., dog, homeless person, coyote) in comparison to every other entity. Each of the 28 dichotomous choice items were presented independently, and the survey automatically progressed after the respondent made their choice. Each of the dichotomous choice items had the same directions, 'Click on the one you would prefer to see in this park' (e.g., Figure 2, below).

The fourth section included a single item requesting participants to rank the eight entities by preference to see in the park. Including both a series of dichotomous choice items and a single ranking item provided a measure of reliability of preferences. The order of the dichotomous choice items and the choices within the ranking item were randomized to eliminate ordering effects. The final section collected information about respondents' behaviour and demographics. This section included the level of park devotion and demographics.

In order to maximize the validity of survey responses, a series of 28 cognitive interviews were conducted in two waves before survey data collection (Willis, 2005). The results of these interviews led to modifications to reduce the cognitive burden associated with number of required 'clicks', clarification of the described purpose of the survey, alteration of two icons to make them more neutral, and expansion of the gender item to include a non-binary response option. There were no additional consistent issues identified during the first round of in-person survey data collection.

Data collection

Survey

The first convenience sample of survey respondents was collected using an intercept method at four Northern California universities and one airport. Although the authors considered collecting

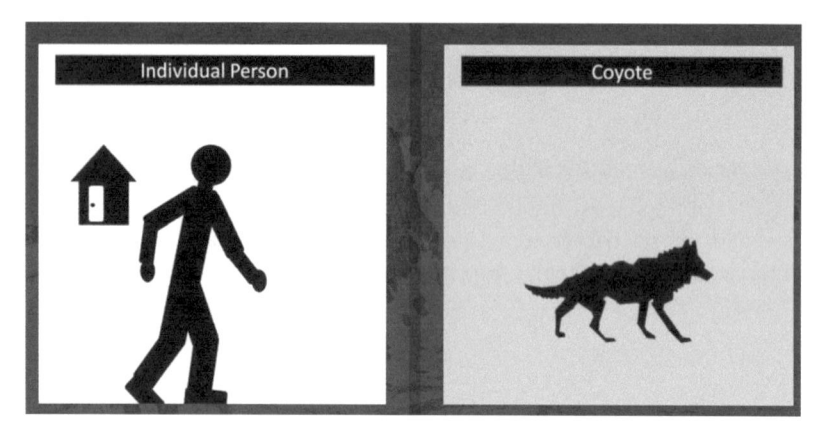

Figure 2. Example of dichotomous choice survey item approximately here.

data at parks, the concern was that this method would oversample more frequent park users. In contrast, the goal of the study was to generalize across type of park and park user, so the choice was made to avoid collecting data at parks. Furthermore, the data was collected in California because that state has had the greatest number of reported coyote incidents (Baker & Timm, 2017). At the intercept points, research assistants asked respondents to take a 10-min survey concerning parks. There was no incentive for participation.

The goal of the second sample was to compare the attitudes of the Californian sample to a more geographically dispersed sample of people living in the US The second sample used the Amazon Turk service to collect responses from people in the US (operationalized as people responding using American IP addresses). The survey was described to potential respondents as a 10-min academic survey focused on parks. Participants were given an incentive of $0.70 USD to participate. The sample of American respondents was scheduled on a Saturday morning, in order to collect a sample that was older and more likely to be employed full-time (Arechar, Kraft-Todd, & Rand, 2017) to complement the first sample which was largely collected on university campuses.

Interviews

Interviewees were purposely selected to maximize participant diversity. In particular, participants were selected to include a balance of male and female participants, dog guardians and non-guardians, and participants of varying ages, immigrant status, and experience with homelessness. During the first round of interviews, the primary author recruited six people to participate in interviews. Those people were known to the author and were chosen to maximize demographic diversity. A team of trained interview assistants recruited and interviewed an additional 22 people in the second round of interviewing. Interviews either conducted through Zoom, a video calling service that created video and audio recordings and initial transcriptions, or in-person using a hand-held audio recording device. Most interviewees participated in the interview in their residence; however, some participated in a public place (e.g., café or place of work). The interviews included the think-aloud method, where participants explicated their reasoning and thought process as they responded to the electronic survey (Presser et al., 2004; Willis, 2005). Although interviewees responded to the survey, none of their responses to the survey were included in the analysis of the survey data. Moreover, interviewers used verbal probes, both extant probes inquiring why participants responded to an item in the way that they did as well as emergent probes further clarifying or gaining additional information about their initial responses.

Data analysis

Survey data were downloaded from Qualtrics into Excel. Data were cleaned, coded, and exploratory analyses were conducted in Excel. Subsequently, data were uploaded into IBM SPSS v.24. where t-Tests, ANOVAs, Friedman tests, and Wilcox signed rank tests were performed.

Data from the interviews were used to both modify the survey instrument and qualitatively contextualize the results of the study. Interview transcriptions were coded according to first in relation to the eight entities (e.g., coyote, off-leash dog, person experiencing homelessness) then subsequently using emergent codes (Brinkmann, 2014). The quotes with their extant entity and emergent codes were compared against each other to gain additional insights about the data and facilitate further development of emergent codes. Transcripts were then re-read in comparison to the emergent codes in order to continue the hermeneutic circle and further refine the codes (Kvale, 1983).

Results

Results are provided for both survey and interview data. Survey responses ($n= 482$) indicate that coyotes were consistently ranked as the least preferred option for sharing park space with people. Data from 28 interviews were coded into four themes; safety, (out of) place, affinity for wildlife, and conspecific loyalty.

Participants

Survey

A total of 510 responses were initially collected (first round $n= 287$, second round $n= 223$). Eight responses were deleted because the respondent failed to correctly respond to the two attention testing items (first round:5, second round:3). An additional 20 responses were deleted because the respondents failed to complete two or more dichotomous choice items or did not complete the single ranking item (first round:13, second round:7). A third attention check involved the 28 dichotomous choice survey items. No respondents clicked only the images that appeared on the left, or only on the images that appeared on the right, so this attention check led to no further removal of responses. This resulted in 482 useable responses (first round $n= 269$, second round $n= 213$).

As intended, the second sample was more geographically diverse and older than the first sample (Table 1). Nearly all (95%) of the primary parks (park they most frequently spent time in for 15 min or more) in the first sample were in California. In contrast, the online sample included primary parks in 43 different states and only 8% of the primary parks in the second sample were in California. Moreover, online respondents communicated a higher rate of experiencing homelessness, dog guardianship, and walking their dog in a park. In contrast, there were no statistically significant differences between respondents' gender, level of education, or household income.

Table 1. Comparison of demographics of survey respondents.

Demographic variables	Sample 1 ($n= 269$)	Sample 2 ($n= 213$)	p-value
Gender (*Female*)	52%	47%	.227
Age (*35 years or older*)	23%	48%	<.001***
Education (*Bachelor's or higher*)	52%	54%	.621
Household Income (*$50k or more*)	47%	46%	.878
Homelessness (*past or current*)	9%	18%	.004**
Dog guardian	45%	54%	.038*
Regularly walks dog in park	57%	74%	.006**

*$p< .05$
**$p< .01$
***$p< .001$

Interviews

Twenty-eight interviews were conducted. Interview recording durations ranged from 14 min 37 s to 53 min 26 s ($\bar{x}$= 30 min 41 s), resulting in more than 14 h of recordings. Gender was nearly evenly balanced between women (n =15) and men (n= 13). The sample was relatively young. Over half (n= 15, 54%) of the participants were in their twenties. Over two-thirds (n= 20, 71%) of the sample was White. One-fifth (n= 5, 18%) of the sample were immigrants to the US (e.g., Europe, Asia, or Latin America). Three (11%) of the interview participants had previously experienced homelessness. Eleven participants (39%) currently owned a dog. An additional participant's dog had recently died, but she still considered herself a dog guardian. The number of dogs owned by any participant varied between one to three dogs.

Ranking

Ranking for preference to share park space with the eight entities; person, person visibly experiencing homelessness, person with on-leash dog, person with off-leash dog, homeless person with on-leash dog, homeless person with off-leash dog, dog, and coyote; ranged from 1 *most preferred* to 8 *least preferred*. The median rank for coyotes in both samples was 8 *least preferred* (Table 2, Figure 3).

Analysis of the single ranking item indicated that the coyote option was the least preferred option for both the first (Friedman mean rank = 6.58, $X^2(7)$ = 895.529, p< .001) and second sample (Friedman mean rank = 7.05, $X^2(7)$ = 734.700, p< .001). The analysis of the dichotomous items similarly identified that 'coyote' was the lowest ranked entity in the first (Friedman mean rank = 6.45, $X^2(7)$ = 895.529, p< .001) and the second sample (Friedman mean rank = 6.78, $X^2(7)$ = 734.700, p< .001). This provides reliability evidence that the rankings of preference for sharing public park spaces with coyotes were consistent across samples and question form (dichotomous choice items or ranking item).

Table 2. Coyote ranking descriptive statistics.

	Sample 1	Sample 2
N	269	213
Mean	6.58	7.05
Median	8.00	8.00
Std. Deviation	2.392	1.914
Skewness	−1.482	−2.023

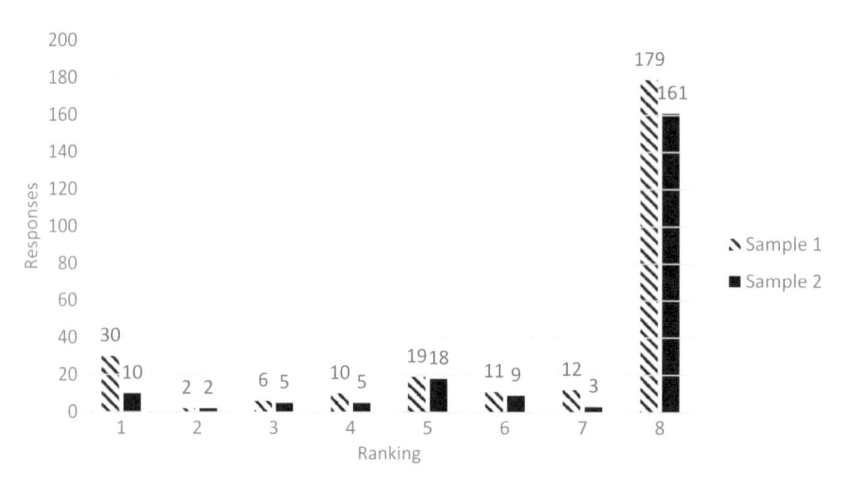

Figure 3. Distribution of coyote rankings.

The mean rank of coyotes was higher (i.e., less preferred) for each demographic, behavioural, or attitudinal subgroup with some exceptions. Sample 1 respondents that were 35 years or older (n= 60) ranked both the off-leash dog ($\bar{x}$= 6.43) and person experiencing homelessness with an off-leash dog ($\bar{x}$= 6.13) as less preferred than a coyote ($\bar{x}$= 5.95). Moreover, sample 1 respondents that had experienced homelessness (n= 23) ranked the off-leash dog ($\bar{x}$= 6.35) as less preferred than the coyote ($\bar{x}$= 6.13).

The ranking of coyotes did not significantly differ by most demographic, behavioural, or attitudinal subgroups (Table 3). Males and respondents that visited parks and other green spaces at least once per week significantly (p< .05) preferred coyotes (closer to 1) compared to female respondents, and respondents that visited other parks and green spaces less frequently.

Description of preferences

To better understand some of the nuances and contextual factors that contribute to understanding preferences for coyotes in urban parks, interviews with survey respondents elicited narratives of meaning and understanding. Data from the interviews were coded into four emergent themes: *safety, (out of) place, affinity for wildlife, and conspecific loyalty* (Table 4).

Safety

The most common issue associated with participants' descriptions of their coyote ranking in the survey was safety. Participants discussed their fear that coyotes might harm their pets, other people, or themselves. The number of coyotes and the participant's experience with coyotes were associated with different levels of fear.

Table 3. Coyote ranking by subgroup.

Variables	Sample 1 $\bar{x}$(sd)	p-value $p(\eta^2)$	Sample 2 $\bar{x}$(sd)	p-value $p(\eta^2)$
Gender				
Male	6.25(2.574)	.018*(.021)	6.82(2.152)	.032*(.022)
Female	6.94(2.131)		7.37(1.447)	
Household income				
Less than $50k	6.48(2.453)	.509	7.24(1.867)	.113
$50k or more	6.68(2.388)		6.83(1.953)	
Education				
Less than a bachelor's degree	6.92(2.192)	.041*(.016)	7.05(1.992)	.997
A bachelor's degree or greater	6.32(2.515)		7.05(1.854)	
Age				
Less than 35 years	6.79(2.245)	.016*(.022)	7.12(1.788)	.604
35 years or more	5.95(2.746)		6.98(2.049)	
Homeless experience				
No	6.64(2.338)	.325	7.28(1.610)	.000***(.066)
Yes	6.13(2.881)		6.00(2.721)	
Frequency visiting primary park				
Fewer than 1 visit per week	6.64(2.375)	.690	7.20(1.928)	.313
1 or more visits per week	6.52(2.424)		6.93(1.914)	
Attachment to primary park				
Neither, Somewhat, or Extremely Unattached	6.77(2.181)	.389	6.65(2.203)	.110
Extremely or Somewhat Attached	6.50(2.503)		7.16(1.818)	
Frequency visiting other parks and green spaces				
Fewer than 1 visit per week	6.82(2.208)	.034*(.017)	7.20(1.770)	.009**(.032)
1 or more visits per week	6.17(2.657)		6.29(2.408)	
Dog guardian				
No	6.88(2.077)	.036*(.017)	7.09(1.932)	.775
Yes	6.26(2.692)		7.02(1.906)	
Regularly walks dog in park				
No	6.37(2.690)	.696	7.48(1.455)	.175
Yes	6.18(2.710)		6.95(1.906)	

Table 4. Interview themes & subthemes.

Theme	Subthemes
Safety	Fear & danger
	Experience with coyotes
	Other users' safety
	Pets
	Single versus multiple coyotes
(Out of) Place	Wild creature in a domesticated space
	Wandering coyote
	Home in the park
	Novelty & excitement
Affinity for wildlife	Love of nature
	Ecological balance
Conspecific loyalty	Unsheltered homelessness vis-à-vis coyotes

Fear & danger

Although one participant admitted he did not know enough to evaluate whether coyotes were dangerous, 'I don't know how dangerous coyotes actually are' (Tyler), most participants expressed some level of concern about coyotes. At one end of the spectrum, John declared, 'The coyote presents a danger in this community and I would want to see Animal Control help take the coyote to a safer location.' In contrast, Robert, an experienced outdoors person stated, 'I have an extremely tiny amount of fear associated with coyotes. I've never experienced a problem, so why should I?'

The most common reasoning for not fearing coyotes was that they are more scared of humans than we should be of them. 'This is an interesting question because coyotes don't freak me out, because they're usually scared of you' (Mike).

Experience with coyotes

Robert, Mike, and Santiago directly or indirectly referenced their rural experiences as the basis for their lack of fear. 'Maybe just because I'm a country guy and I like, kind of, I like seeing the coyotes, they don't scare me or anything' (Mike). Santiago stated, 'I grew up in the mountains of Mexico and later moved out in the country of the States. I have heard and seen a lot of coyotes. They didn't bother me. I didn't bother them.' Ellie contrasted her own fear and inexperience with coyotes with other people's experience, 'If they've seen coyotes, they're comfortable with coyotes. They'd probably be like, "Screw people, I'm comfortable with coyotes."' Terry also expressed her perception that coyotes were scared of humans, though she was adamant that she did not want a coyote in her park.

Other users' safety

Even though multiple participants claimed not to be concerned about their safety relative to coyotes, they addressed the issue of safety without being prompted by interviewers. This prioritization suggests they were aware that other park users may be concerned about their own safety around coyotes. For example, even though she thought that she would not be bothered by a coyote, Sofia said she would take steps to minimize the chance it would hurt other park visitors.

> I would inform the people that work at the park if there was a coyote. People need to be aware of the coyote.
> You don't want it to harm anyone and keep it to yourself. You worry about the rest of the people at the park.

Participants often perceived that children were the most vulnerable population to coyotes. 'Coyotes in a park seems a little dangerous... especially at ... there's a lot of kids' (Corey). Shiva similarly communicated his concern about children, 'I am worried about the safety of the kids who come out to play. I don't want my kid to come in contact with a homeless (person) or his dog or coyote.' Tyler stated, 'I'd be kind of worried about my kids going to a park that has a bunch of coyotes, but I think it'd be fine seeing a coyote.' Corey graphically described this concern, 'A coyote might try to bite me, I dunno... eat some babies, I dunno!'

Pets

Beyond human children, multiple people shared their concern that coyotes might hurt their or other people's pets. Terry shared, 'I grew up next to a canyon with coyotes and they killed many dogs in the neighborhood over the years.' Katrina, the guardian of two small dogs, explained her desire to not see coyotes by flatly declaring, 'Coyotes eat dogs.'

Comparisons of coyotes with dogs clarified the relative perceived danger of coyotes versus off-leash dogs with no humans. Many of the interview participants compared the coyote with the off-leash dog with no people and concluded that, as expressed by Elena, the dog was 'less risk than a coyote.' Alanis claimed that, 'Dogs are usually more friendly than coyotes.' She subsequently further described her concern with coyotes, 'I get dogs off a leash are scary, but they aren't trying to hunt you.' Nathan stated, 'I don't want to come across a coyote in the park and I can still manage with an off-leash dog since I am a dog owner myself and know the dog psychology a little.' Shiva had a similar claim that he was more prepared to deal with an off-leash dog, 'I know (a) little dog language and probably can do away with a dog on (the) loose... But a coyote? I still want to live!' Petra originally claimed that, 'I'm not bothered by them. My dog doesn't react to them in a negative way,' but she also thought that a coyote was more dangerous than an off-leash dog with no associated human.

The perception of the off-leash dog with no humans was sometimes associated with a back-story participants created for the dog. Petra expressed that she thought the dog might be lost and need help. Similarly, Sofia thought that the dog was just separated from its guardian and it likely needed food. 'I know that the dog is probably hungry and he is with his guardian. The dog is less likely to harm me than a coyote.'

Single versus multiple coyotes

One dimension of danger mentioned by three participants was the number of coyotes seen at one time. For example, Santiago expressed, 'I am actually okay with seeing a coyote. I would not feel threatened unless it's a pack.' In contrast, Robert was eager to seem multiple coyotes, 'I'd prefer to see a pack of coyotes over a single coyote.'

(Out of) place

There were substantial comments about the juxtaposition of wild coyotes in domesticated spaces and whether a coyote belonged in that space. Although the previously discussed issues around fear were present for some respondents, others discussed the novelty and excitement of seeing coyotes in their urban or peri-urban parks.

Wild creature in a domesticated space

One of the dichotomies that participants alluded to when making the argument that coyotes should be excluded from urban parks was the concept of wildness. It was reasoned that *wild* coyotes should be excluded from *domesticated* park spaces. '(A coyote) is a wild animal, and if it was just walking around by itself in a park, it would seem very out of place' (Anna). Nathan referenced the same concept of wildness, 'I would never want to see a coyote in the park. It is like getting a wild animal into an urban city park.' Similarly, Shiva exclaimed, 'I want to see a coyote on a safari... not in my park where I am on a walk.'

Wandering coyote

There were multiple expressions of disbelief that a coyote would ever be in an urban park, 'Tell me something... what's up with the coyotes? Is someone really trying to get coyotes in the park? '

(Nathan). Jeremy thought that a coyote could only be at his suburban park if it was lost. 'Seeing a coyote in this area would be out of the norm... I would also be worried about the coyote it most likely wandered into this area by accident and is lost.' John similarly claimed that seeing a coyote in a metropolitan area would be abnormal. 'I live in a suburban area that is right outside of a metropolitan area. Seeing a coyote out here is by no means normal.'

Home in the park

Multiple participants expressed the idea that they felt comfortable with seeing coyotes in the park because that was a normative space for coyotes. There were two general concepts that these participants referenced, 'what is normal' and 'home'. Ginny expressed that in her experience, 'I feel like that would be a normal thing to see in a park.' Santiago explained, 'I think it's pretty okay to see a coyote out on the trail. It's their home, we are in their territory. Wildlife was there first.' Linnea referenced this concept succinctly by stating, 'Because they belong there and we don't.' Ginny explained that coyotes were appropriate in natural places because they had relatively fewer options than dogs and people, 'You can walk your on-leash dog in a lot of different places. But a coyote can only be in so many.'

Novelty & excitement

While many participants did not want to see a coyote in their primary park (e.g., 'Coyotes give me the creeps. Why would anyone want to see a coyote in the park?'), multiple participants wanted to see a coyote because it would be a rare or exciting experience to see a coyote in their urban or peri-urban park (e.g., 'I would like to see the coyotes, just because I love seeing the coyotes... You gotta love the coyotes!' <Geraldine>). Charlie described why seeing a coyote in the downtown park he frequented would be the most desirable option:

> That's the only one of these options that would be like, 'Whoa, check it out, there's a coyote here!' Because you're like in the middle of the heart of the city. You're not like in the or anything... That's the only one that would be a story to tell. Like, 'Dudes, I saw a coyote, next to <place of work>!

These responses point to the juxtaposition of seeing a 'wild' animal in a built, urban setting, interconnecting with other subthemes. 'I just think that that's cool. I always love to see coyotes. It is a special treat' (Walter). Carmen struck a more tentative tone than the participants quoted above. 'Most people would be scared of them, but I think I'd be more fascinated to see one in person. I'd still take caution though.'

Affinity for wildlife

Beyond debating whether the coyote should or should not be in their primary park, some participants wanted to see a coyote because it met their interest to experience nature.

Love of nature

Sometimes this was simple, because the participant expressed both a desire to see coyotes and a love of wildlife. 'I think it would be pretty cool to see a coyote there actually, just because I like wildlife' (Mike). Robert explained, '(A coyote) is part of what I am hoping for is a natural experience which is part of why I leave the comforts of town and home.' Similarly, Tyler stated, 'That'd be rad (to see a coyote). When I'm going to the park I'm not there to see people, I'm there for nature anyway.'

Ecological balance

However, there was also conflicts between ideals about wildlife and a desire to see a coyote in their primary park. Nargis' fear of coyotes and lack of desire to see them conflicted with her ideals about wildlife. 'If you feel you have these kind of creatures around in the park, it's kind of healthy balance, right? You still have wild around you.' Similar to Nargis, Javier expressed safety concerns, but he also valued coyotes as free wild animals, "Cause coyotes should live in nature and they should not be in some type of cage or something. I believe that coyotes should also be free.' Moreover, he thought that coyotes needed access to the natural space more than dogs, 'The dogs have been domesticated, so the dogs could be happy just outside your house. And the coyote is something different.'

Conspecific loyalty

The final emergent theme was that other humans should be preferred over other species, including coyotes. Terry expressed, 'Who would choose encountering a coyote over a person experiencing homelessness? I mean I'm sure somebody would…that makes me sad that someone would choose a coyote over someone who is down on their luck.' Similarly, Ginny thought that people should come before animals, 'I guess coyotes have only a limited place to be in a park, but I feel like I think homelessness takes precedence.' Geraldine shared an opposing view, 'I feel that is (coyotes') natural habitat and the homeless people have other choices as to where they can stay.'

Discussion

Survey respondents ranked coyotes as the least preferred entity with which to share public park space. Although the current study is cross-sectional and therefore cannot measure changes in attitudes across time, concern that coyotes are being embraced by modern Americans (e.g., Kellert, 1985; Timm, 2006) is not supported by the current study.

The lack of preference to share park space with coyotes suggests that even among the sometimes marginalized groups focused on in this study (people experiencing homelessness, dogs, and coyotes) coyotes are the least preferred. The qualitative analysis of the interviews suggests that fear for their own and other people's safety, a perception that coyotes do not belong in the city, and a relative preference for other humans all may have contributed to coyotes' relatively poor ranking. Alternatively, the interviews also suggest that a perceived lack of danger, a belief that coyotes have a right to park space, and an affinity for wildlife suggest ways that park managers could support a greater tolerance or appreciation for coyotes in urban and peri-urban parks.

Dangerous creature

Although some participants were eager to share space with a coyote, the survey responses indicate that the majority of respondents preferred to not share public park space with a coyote. The interview data strongly suggest that fear was based on a perception of coyotes as dangerous creatures that would harm themselves or others. Participants feared the supposedly wild nature of these animals (Rose & Wilson, in press).

Alexander and Quinn (2012) analysis of the portrayal of coyotes in Canadian media found that, regardless of the actual hazard posed by coyotes, they were often labelled negatively based on the actions of a few individuals and that coyotes were often described as *bold, brazen,* and *wily criminals* with *malicious* intent. But how dangerous are they to humans and pets?

Attacks on humans

From 1997 to 2015, Baker and Timm (2017) identified 367 coyote attacks on humans. Nearly half of the known attacks occurred in California; however, the geographic range of attacks has increased over time (Baker & Timm, 2017; White & Gehrt, 2009). There are only two known people that have been killed by coyotes, one in California in 1981 and one in Nova Scotia in 2009. Specific to the focus of the current study, White and Gehrt (2009) media analysis found that one-quarter (25%) of the victims were in parks and half (47%) were recreating when they were attacked. Similarly, Alexander and Quinn (2011) media analysis found that 42% of humans were playing or walking in greenspace when they had an interaction with a coyote.

Although coyotes have attacked humans, it remains a rare occurrence. White and Gehrt (2009) calculated that there were 3.46 annual incidents of coyotes biting humans in their North American study, and Alexander and Quinn (2011) identified an average of 2.67 annual bites in their Canadian study. Baker and Timm (2017) more liberal definition of an attack results in an annual average of less than 20 (19.32) attacks annually. To put this in perspective, there are approximately 4.7 million incidents of dogs biting humans each year in the US (Gilchrist, Gotsch, Annest, & Ryan, 2003).

Multiple interview participants shared their concern that coyotes pose an acute danger to youth. This concern is often echoed in media reports and some early studies of coyote attacks (e.g., Carbyn, 1989; Fimrite, 2015). One of the two known incidents of a human being killed by a single or multiple coyotes was a three-year-old girl in Southern California (Howell, 1982). However, an analysis of all attacks indicates that children are no more likely to be attack victims than adults (White & Gehrt, 2009). In contrast, children are more likely to be the victims of dog bites in the US (Gilchrist et al., 2003).

In the current study, men preferred sharing space with coyotes more than women. More than 30 years ago, Kellert (1985) similarly identified that women preferred predators, including coyotes, less than men. However, there is no evidence that women are more likely to be a victim of a coyote attack compared to men (White & Gehrt, 2009).

Attacks on pets

In contrast to the relatively small threat coyotes pose to humans, coyotes can be dangerous neighbours for domestic dogs and cats. Baker and Timm (2017) speculated that the 'coyote attacks on pets in some suburban areas are now so common that they are no longer considered news' (p. 124). The authors reasoned that the number of coyote attacks on pets may be increasing because of a growing number of coyotes and pets, a greater habituation of coyotes, or an increased rate that people report the loss of their pets to authorities (Baker & Timm, 2017). This warning about the increase in number of coyote attacks on pets was matched by the warning that increases in pet predation foretell an increased risk to human safety (Baker & Timm, 1998).

Coyotes eat cats. While some studies have found evidence of consumption of relatively low levels of cats in urban coyote diets (<1%-6.7%; Morey, Gese, & Gehrt, 2007), others studies suggest cat consumption is a more prominent part of urban coyotes' diet (Quinn, 1997; Shargo, 1988). Alexander and Quinn (2011) reported that all of the coyote-cat incidents in their Canadian media analysis resulted in the death of the cat; however, such a media analysis may suffer from a selection bias where only the more extreme cases of domestic animal death were recorded. In their observational study, Grubbs and Krausman (2009a) observed 36 coyote–cat interactions. The cats were killed in 19 of those interactions. The authors concluded that 'any cat outside is vulnerable to coyote attack' (Grubbs & Krausman, 2009a, p. 684).

Dogs are also potential coyote prey. 'Animal Care and Control assures us that coyotes don't really want anything to do with humans, and it's just your dogs, especially small ones, that you should worry about if they're running around off-leash' (Barmann, 2016). In a review of Canadian media from 1995 to 2010, Alexander and Quinn (2011) identified 108 coyote–dog interactions. Nearly two-thirds (63.7%) of the interactions were with small dogs where most (59%) were killed. In contrast, only 1 of the 25 large dogs reported interacting with coyotes died.

Such predation may support the population of songbirds and other potential cat and dog prey species (Crooks & Soulé, 1999; Gehrt, Wilson, Brown, & Anchor, 2013; Gompper, 2002), but ignores the emotional loss of pet guardians. Webster (2007) claimed that, in contrast to wild animals which are valued at population levels, pets are valued as individuals. One such grieving dog guardian referenced this distinction with a statement that demarcated the in-group pets from the out-group wild coyotes, 'Something has got to be done. *Our* animals are in danger' (Batey, 2016).

Therefore, the concern expressed by some interview participants that coyotes were dangerous for their small four-legged family members appears to be a correct assertion. Katrina's assertion that 'Coyotes eat dogs,' should be expanded to 'Coyotes may eat any small mammal they can catch, including small off-leash dogs and free-roaming cats.'

A coyote's place is...

In the interview data, there is a clash between the view that coyotes have a claim on the limited urban green space and those that strongly deny that claim. The natural range expansion of coyotes into all parts of North America and their ability to live in limited natural areas in urban areas means that, in contrast to the perception expressed by many interview participants, the participants' primary parks are already likely part of coyotes' home range (Bounds & Shaw, 1994; Gompper, 2002). The abundance of food in urban areas allows coyotes to maintain smaller home ranges than coyotes in rural areas (Gehrt, Anchor, & White, 2009; Grinder & Krausman, 2001). Therefore, the density of coyotes in urban areas is often greater than in rural areas (Atwood, Weeks, Gehring, & White, 2004; Fedriani, Fuller, & Sauvajot, 2001; Kenaga, Krebs, & Clapham Jr, 2013). In the confined urbanity of San Francisco, it has been estimated that there are over 100 individual resident coyotes (Stienstra, 2014).

Not everyone is receptive to this current reality. These conflicts with 'the idea that cities are the exclusive domain of humans' (Wolch, 2002, p. 726). In her essay against coexistence with coyotes, Webster (2007) stated, 'Cities are not for the Third Worldness of the Wild Kingdom' (p. 107). This perspective sees coyotes as an affront to attempts at eliminating the *pollution* of animals from anthropocentrically-focused sanitized urban environment (Jerolmack, 2008; Wolch, 2002). Coyotes seem 'doomed to be considered morally transgressive as they transgress the spaces we have defined as "for humans only"' (Jerolmack, 2008, p. 88). They possess a level of agency, which may conflict with some human sensibilities. 'When animal and human trajectories collide in the built environment, to the extent that animals cannot be tamed or controlled, there is an underlying existential human experience of social disorder' (Jerolmack, 2008, p. 89).

Even though coyotes live in urban and peri-urban areas, and make use of parks in these spaces, they generally avoid humans. In contrast to coyotes living in more rural locations who are primarily crepuscular (Bekoff & Gese, 2003), coyotes living in urban and peri-urban areas are nocturnal (Atwood et al., 2004; Gehrt et al., 2009; Grinder & Krausman, 2001; Grubbs & Krausman, 2009b; Moll et al., 2018). This temporal shift is thought to be one way coyotes avoid humans. Moreover, coyotes avoid areas with high levels of human activity (Atwood et al., 2004; Gehrt et al., 2009; George & Crooks, 2006; Mueller, Drake, & Allen, 2018; Reed & Merenlender, 2011). Furthermore, neither resident nor transient coyotes are attracted to areas associated with human activity (Gehrt et al., 2009).

Management implications

Although mass killings of coyotes have been a common practice in North America (Alexander & Quinn, 2011; Howell, 1982; Timm & Baker, 2007), killing coyotes in urban areas has become an increasingly unpopular management action over the last 30 years (Kellert, 1985; Loring, 2015; Vaske & Needham, 2007). Moreover, relocating coyotes is often not an option. In California, where the largest portion of coyote attacks have occurred, it is illegal to relocate coyotes, and lethal measures are used only as the last resort (Batey, 2014). The most common contemporary management strategy appears to be some version of coexisting with coyotes that focuses on educating people to minimize negative interactions (Fimrite, 2015; Timm & Baker, 2007; Webster, 2007).

Although some individuals may desire to see coyotes and may believe that coyotes have a right to public space, survey data suggest that most individuals would rather not have to deal with these creatures in their public parks. The data communicate that there is still a strong sentiment that parks, especially urban parks, are the domain of people and their domestic pets. The pervasiveness of this perspective suggests that authorities either need to engage in some targeted educational campaigns to change people's perceptions of coyotes, modify their management strategy, or risk a backlash of public opinion. The qualitative data from this study suggest that if a strategy of coexistence with coyotes is going to be successful, then management attempts need to directly address humans' fear, perception of coyotes as intruders, and support human's affinity for coyotes and other non-humans.

Education campaigns may start by increasing people's knowledge. Coyotes are not very dangerous to humans. Furthermore, attempts should be made to reduce habituation by eliminating opportunities for coyotes to acquire anthropogenic sources of food (e.g., keep cats indoors, leash dogs, remove fruit and nuts from trees; Alexander & Quinn, 2011). Furthermore, people need education concerning what to do when they encounter a coyote. Residents are often advised to haze coyotes with loud noises, flashing lights, and other aggressive actions (Kessler, 2016); however, any management actions need to be monitored for long-term effectiveness (Baker & Timm, 2017; Bounds & Shaw, 1994). Second, education campaigns may only lead to behaviour change if they first convince people that coyotes are permanent residents in urban and peri-urban areas and extermination or relocation are not viable options. Increasing awareness of the number of coyotes in the area may be one way to support this outcome.

Third, general affinity for wildlife should be supported. This support for a local environmental ethic may be helpful for urban wildlife, and may also promote a more general positive environmental ethic. An important aspect of this goal is that any attempts to support an affinity for wildlife in an urban environment in an urban setting must work with the current perception of some human urban residents that access to public space may be a zero-sum game where support for a non-human animal, such as a coyote, may equate to a loss for the most marginalized human individuals.

Limitations

The sequential mixed method study offers many methodological advantages; however, there are limitations to the study. Although the sampling strategy benefits from samples at the regional (Northern California) and national level, neither are simple random samples, limiting the generalizability of the findings. Although there may be concern that using an online survey service that uses a small incentive for participation such as Amazon Turk could lead to a distinct sample, the major demographic differences were the intended ones, the online sample was older and more geographically dispersed, and the ranking responses of the two samples showed similarities. Furthermore, the survey results did not control for differences in rural versus urban background, nor was the degree of urbanity of the respondents' primary park measured.

Future studies can build on the findings in this study by measuring how respondents' perspectives differ based on the location of the park and the respondent's experience with coyotes and nature more generally. Furthermore, additional research should seek to measure how management strategies can impact people's fear of coyotes, perception of coyote's right to urban space, the public's affinity for nature and how those changes impact the humans' relative preference for sharing park space with coyotes and other wildlife.

Conclusion

Parks are materially constructed spaces that are often socially constructed as spaces of leisure for humans and their canine companions at the same time as they are habitat which other species use to sustain themselves. Evaluations of preference for interactions between different species of park visitors and residents require a reconsideration of which aspects of parks should be considered timeless, and which aspects which should be modified based on contemporary perspectives. This study provides evidence that most people prefer not to share their urban and peri-urban parks with coyotes. If that option is no longer available and coexistence is to be the status quo into the future, then authorities need to engage in educational campaigns to change the perception of predators in urban and peri-urban public spaces.

Disclosure statement

No potential conflict of interest was reported by the authors.

ORCID

Jackson Wilson ⅈD http://orcid.org/0000-0003-1257-9497
Jeff Rose ⅈD http://orcid.org/0000-0003-3171-7242

References

Alexander, S. M., & Quinn, M. S. (2011). Coyote (Canis latrans) interactions with humans and pets reported in the Canadian print media (1995–2010). *Human Dimensions of Wildlife*, *16*(5), 345–359.

Alexander, S. M., & Quinn, M. S. (2012). Portrayal of interactions between humans and coyotes (Canis latrans): Content analysis of Canadian print media (1998–2010). *Cities and the Environment (CATE)*, *4*(1), 9.

Arechar, A., Kraft-Todd, G. T., & Rand, D. G. (2017). Turking overtime: How participant characteristics and behavior vary over time and day on Amazon mechanical turk. *Social Science Research Network*. doi:10.2139/ssrn.2836946

Atwood, T. C., Weeks, H. P., Gehring, T. M., & White, J. (2004). Spatial ecology of coyotes along a suburban-to-rural gradient. *Journal of Wildlife Management*, *68*(4), 1000–1009.

Baker, R. O., & Timm, R. M. (1998). Management of conflicts between urban coyotes and humans in southern California. *Hopland Research & Extension Center*.

Baker, R. O., & Timm, R. M. (2017). Coyote attacks on humans, 1970–2015: Implications for reducing the risks. *Human-Wildlife Interactions*, *11*(2), 120.

Barmann, J. (2016, January 12). Wild coyotes now roaming Haight, Panhandle. Retrieved from http://sfist.com/2016/01/12/wild_coyotes_now_roaming_haight_pan.php

Batey, E. (2014, June 27). How much of a threat are San Francisco coyote's? Retrieved from http://sfist.com/2014/06/27/stay_frosty_san_franciscos_populati.php

Batey, E. (2016, March 18). Fatal attacks spur city hall meeting on SF's coyote response plan. Retrieved from http://sfist.com/2016/03/18/are_coyote_attacks_on_the.php

Bekoff, M., & Gese, E. M. (2003). Coyote (Canis latrans). In G. A. Feldhamer, B. C. Thompson, & J. A. Chapman (Eds.), *Wild mammals of North America: Biology, management, and conservation* (2nd ed., pp. 467–481). Baltimore, MD: John Hopkins University Press.

Bounds, D. L., & Shaw, W. W. (1994). Managing coyotes in US National Parks: Human-coyote interactions. *Natural Areas Journal*, *14*(4), 280–284.

Brinkmann, S. (2014). Unstructured and semi-structured. In P. Leavy (Ed.), *The Oxford handbook of qualitative research* (pp. 277–299). Oxford: Oxford University Press.

Carbyn, L. N. (1989). Coyote attacks on children in western North America. *Wildlife Society Bulletin (1973–2006)*, *17*(4), 444–446.

Creswell, J. W., & Creswell, J. D. (2018). *Research design: Qualitative, quantitative, and mixed methods approaches* (5th ed.). Thousand Oaks, CA: Sage Publications, Inc.

Crooks, K. R., & Soulé, M. E. (1999). Mesopredator release and avifaunal extinctions in a fragmented system. *Nature*, *400*(6744), 563–566.

Fedriani, J. M., Fuller, T. K., & Sauvajot, R. M. (2001). Does availability of anthropogenic food enhance densities of omnivorous mammals? An example with coyotes in southern California. *Ecography*, *24*(3), 325–331.

Fimrite, P. (2015, September 17). S.F. neighborhood on edge after coyote pack moves in, kills cats. *SF Chronicle*. Retrieved from http://www.sfchronicle.com/bayarea/article/S-F-neighborhood-on-edge-after-coyote-pack-moves-6512264.php

Galante, G., (photographer). (2018). *Urban coyote*

Gehrt, S. D. (2004). Ecology and management of striped skunks, raccoons, and coyotes in urban landscapes. In N. Fascione, A. Delach, & M. Smith (Eds.), *People and predators: From conflict to coexistence* (pp. 81–104). Washington DC: Island Press.

Gehrt, S. D., Anchor, C., & White, L. A. (2009). Home range and landscape use of coyotes in a metropolitan landscape: Conflict or coexistence? *Journal of Mammalogy, 90*(5), 1045–1057.

Gehrt, S. D., Wilson, E. C., Brown, J. L., & Anchor, C. (2013). Population ecology of free-roaming cats and interference competition by coyotes in urban parks. *PloS one, 8*(9), 1.

George, S. L., & Crooks, K. R. (2006). Recreation and large mammal activity in an urban nature reserve. *Biological Conservation, 133*(1), 107–117.

Gilchrist, J., Gotsch, K., Annest, J. L., & Ryan, G. (2003). Nonfatal dog bite: Related injuries treated in hospital emergency departments-United States 2001. *Morbidity and Mortality Weekly Report, 56*(26), 605–610.

Gompper, M. E. (2002). Top carnivores in the suburbs? Ecological and conservation issues raised by colonization of North eastern North America by coyotes. *BioScience, 52*(2), 185–190.

Grinder, M. I., & Krausman, P. R. (2001). Home range, habitat use, and nocturnal activity of coyotes in an urban environment. *The Journal of Wildlife Management, 65*(4), 887–898.

Grubbs, S. E., & Krausman, P. R. (2009a). Observations of coyote–Cat interactions. *Journal of Wildlife Management, 73*(5), 683–685.

Grubbs, S. E., & Krausman, P. R. (2009b). Use of urban landscape by coyotes. *Southwestern Naturalist, 54*(1), 1–12.

Howell, R. G. (1982). *The urban coyote problem in Los Angeles County*. Paper presented at the Tenth Vertebrate Pest Conference, Davis, CA.

Jerolmack, C. (2008). How pigeons became rats: The cultural-spatial logic of problem animals. *Social Problems, 55*(1), 72–94.

Kellert, S. R. (1985). Public perceptions of predators, particularly the wolf and coyote. *Biological Conservation, 31*(2), 167–189.

Kenaga, B. A., Krebs, R. A., & Clapham Jr, W. B. (2013). Coyote land use inside and outside urban parks. *American Midland Naturalist, 170*(2), 298–310.

Kessler, J. (2016). Coyote yipps: A blog about San Francisco coyotes. Retrieved from https://coyoteyipps.com/

Kvale, S. (1983). The qualitative research interview: A phenomenological and a hermeneutical mode of understanding. *Journal of Phenomenological Psychology, 14*(1–2), 171–196.

Loring, K. (2015, February 23). In San Francisco, coyotes are your wildest neighbors. *KALW Local Public Radio*. Retrieved from http://kalw.org/post/san-francisco-coyotes-are-your-wildest-neighbors

Moll, R. J., Cepek, J. D., Lorch, P. D., Dennis, P. M., Robison, T., Millspaugh, J. J., & Montgomery, R. A. (2018). Humans and urban development mediate the sympatry of competing carnivores. *Urban Ecosystems, 21*, 765–778.

Morey, P. S., Gese, E. M., & Gehrt, S. (2007). Spatial and temporal variation in the diet of coyotes in the Chicago metropolitan area. *American Midland Naturalist, 158*(1), 147–161.

Mueller, M. A., Drake, D., & Allen, M. L. (2018). Coexistence of coyotes (Canis latrans) and red foxes (Vulpes vulpes) in an urban landscape. *PloS one, 13*(1), e0190971.

Presser, S., Couper, M. P., Lessler, J. T., Martin, E., Martin, J., Rothgeb, J. M., & Singer, E. (2004). Methods for testing and evaluating survey questions. *Public Opinion Quarterly, 68*(1), 109–130.

Proctor, J. D. (1998). The social construction of nature: Relativist accusations, pragmatist and critical realist responses. *Annals of the Association of American Geographers, 88*(3), 352–376.

Quinn, T. (1997). Coyote (Canis latrans) food habits in three urban habitat types of western Washington.

Reed, S. E., & Merenlender, A. M. (2011). Effects of management of domestic dogs and recreation on carnivores in protected areas in northern California. *Conservation Biology, 25*(3), 504–513.

Rose, J. (2017). Cleansing public nature: Landscapes of homelessness, health, and displacement. *Journal of Political Ecology, 24*, 11–23.

Rose, J., & Wilson, J. (in press). Assembling homelessness: A posthumanist political ecology approach to urban parks, nature, wildlife, and actor-networks. *Leisure Sciences*. doi:10.1080/02614367.2019.1586979

Rutherford, S. (2018). The Anthropocene's animal? Coywolves as feral cotravelers. *Environment and Planning E: Nature and Space, 1*(1–2), 206–223.

Shargo, E. S. (1988). *Home range, movements, andactivity patterns of coyotes (Canis latrans) in LosAngeles suburbs* (Ph.D. Dissertation). UCLA, Los Angeles.

Stienstra, T. (2014, March 27). Coyotes seemingly thrive in San Francisco. Retrieved from http://www.sfgate.com/outdoors/article/Coyotes-seemingly-thrive-in-San-Francisco-5045034.php

Taylor, D. E. (1999). Central Park as a model for social control: Urban parks, social class and leisure behavior in nineteenth-century America. *Journal of Leisure Research, 31*(4), 420–477.

Timm, R. M. (2006). *Coyotes nipping at our heels: A new suburban dilemma*. Paper presented at the Triennial Naiotnal Wildlife & Fisheries Extension Specialists Conference, Big Sky, MT.

Timm, R. M., & Baker, R. O. (2007). *A history of urban coyote problems.* Paper presented at the Wildlife Damage Management Conferences, Corpus Christi, TX.

Vaske, J. J., & Needham, M. D. (2007). Segmenting public beliefs about conflict with coyotes in an urban recreation setting. *Journal of Park and Recreation Administration, 25*(4), 79–98.

Webster, J. C. (2007). *Missing cats, stray coyotes: One citizen's perspective.* Paper presented at the Wildlife Damage Management Conferences, Corpus Christi, TX.

White, L. A., & Gehrt, S. D. (2009). Coyote attacks on humans in the United States and Canada. *Human Dimensions of Wildlife, 14*(6), 419–432.

Willis, G. B. (2005). *Cognitive interviewing: A tool for improving questionnaire design.* Thousand Oaks, CA: Sage Publications.

Wilson, J. D., Yoshino, A., & Latkova, P. (2018). Off-leash recreation in an urban national recreation area: Conflict between domesticated dogs, wildlife, and semi-domesticated humans. In J. Young & N. Carr (Eds.), *Domestic animals, humans, and leisure* (pp. 63–81). New York: Routledge.

Wolch, J. (2002). Anima urbis. *Progress in Human Geography, 26*(6), 721–742.

Index